LIFE
AT THE
EXTREMES

LIFE
AT THE
EXTREMES

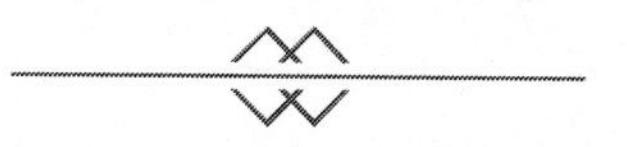

Frances Ashcroft

University of California Press
Berkeley • Los Angeles

University of California Press
Berkeley and Los Angeles, California

Published by arrangement with
HarperCollins*Publishers, Ltd.*

Library of Congress Cataloging-in-Publication Data

Ashcroft, Frances M.
Life at the extremes / Frances Ashcroft.
p. cm.
Published simultaneously in Great Britain by HarperCollins
Includes bibliographical references and index.
ISBN 0-520-22234-2 (cloth : alk. paper)
1. Adaptation (Physiology) 2. Extreme environments. I. Title.
QP82 .A75 2000
612'.014--dc21

00-028672

Set in Giovanni

Printed and bound by
Clays Ltd, St Ives plc

9 8 7 6 5 4 3 2 1

We shall not cease from exploration
And the end of all our exploring
Will be to arrive where we started
And know the place for the first time.
T.S. ELIOT, 'Little Gidding'

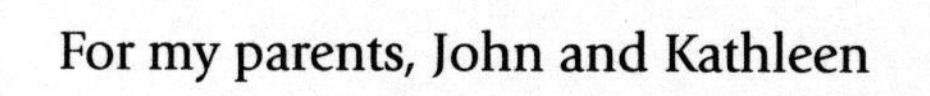
For my parents, John and Kathleen

CONTENTS

LIST OF ILLUSTRATIONS

5. LIFE IN THE FAST LANE

6. THE FINAL FRONTIER

7. THE OUTER EDGES

ACKNOWLEDGEMENTS

This book came about in a most unusual way. In 1998, the Wellcome Trust offered a prize to enable a practising life scientist to take time off from research to write a science book for the general reader. Although I had no intention of taking time off (because my research is far too interesting to stop), I had always been interested in writing and the competition was just the incentive I needed. I talked about it incessantly to friends and colleagues, and agonized over the choice of topic. Months passed. Three weeks before the deadline, I still had not written anything – there were too many interesting subjects to choose from and so little spare time. And then one night I stopped at a friend's house on my way home. To my surprise, she showed me her completed entry for the Wellcome Trust prize (an outline of the book and a specimen chapter), explaining that my enthusiasm had inspired her to enter. I was speechless – and galvanized into action. I started my entry as soon as I got home, choosing as a subject the adaptations that enable humans to survive in extreme environments, the area of physiology that I happened to be teaching in at the time. I did not, in the end, win the prize, but I was lucky for Philip Gwyn-Jones and Toby Mundy nevertheless commissioned me to write a book. This is the result.

I could not have written the book without a great deal of support. I am grateful to many people for reading the chapters and for ensuring my facts are accurate. My parents, my brother Charles, Fiona Gribble and Stefan Trapp bravely tackled them all. Many other people read individual chapters, or parts of chapters, and gave invaluable comments on content and style. I thank Judy Armitage, Hilary Brown, John Clarke, Jonathan Deane, Keith Dorrington, Clive Ellory, Don and Mary Gribble, Abe Guz, Albert Harrison, Michael Horsley, Sally Krasne, Ann Lingard, Phillippa Jones, Cathy Morriss, David Paterson, Peter Robbins,

David Rogers, Janet and Ken Storey, Zbigniew Szydlo, Michael Vickers, Martin Wells, Graham Wilson and Gary Yellen. I was also helped in various ways by many other people. Sandra Moony, David Flowers and David Irvine at British Airways were kind enough to give me several hours of their time to explain about aviation medicine, and both they and David Bartlett provided valuable information on the effects of cosmic radiation. Roger Black answered my numerous naive questions on athletics. Edith Hall supplied me with references to early Greek divers; Gildas Loussouarn helped me translate various French texts; Laurence Waters assisted with the photographs. Justin Wark patiently answered my questions about basic physics; and Judy Armitage corrected my microbiological misconceptions. Hilary Brown, Keith Dorrington, Abe Guz, Michael Horsley, David Paterson, Peter Robbins, Janet and Ken Storey, and Martin Wells helped ensure that the physiology is correct. My mother sent me a stream of relevant newspaper cuttings and my brother was a fund of interesting stories. I thank them all. As Isabel Allende said in the preface to *Aphrodite*: 'to copy one author is plagiarism, to copy many is research.' I also thank my many sources for information and inspiration – and, occasionally, for the loan of a particular turn of phrase that encapsulates something perfectly.

I owe a very special thank you to Peter Atkins, who ensured that I finished the book by telling me I'd never have the discipline to do so! (He knew I would rise to the challenge.) Sarah Randolph provided encouragement when the writing seemed a never-ending process, as well as the spur to start the task in the first place. I also thank the Wellcome Trust for inspiring the life science community to write books for the general reader, and, of course, for funding my scientific research.

I could never have written this book without the help of Jenny Griffiths who photocopied articles endlessly, and winkled obscure books out of the depths of the Bodleian Library. Cathy Morriss helped with the cover design, Suzanne Collins did a marvellous job finding the illustrations, Terence Caven designed the book and Janet Law was a great copy-editor. Most of of all, I thank my wonderful editors, Philip Gwyn Jones and Georgina Laycock at HarperCollins and Howard Boyer at the University of California Press, for their constant support, encouragement and wise advice.

INTRODUCTION

In November 1999, the newspapers were dominated by stories of the death of the golf champion Payne Stewart and four others in an air crash. Their Lear jet lost contact with ground control soon after taking off from Orlando, Florida, at an altitude of around 11,300 metres (37,000 feet). Concerned that it might crash in a populated area, US officials scrambled two Air Force fighter planes – to shoot it down, if necessary. They reported that there was no sign of life on board the Lear jet and that the windows were frosted over, which suggests that the aircraft had depressurized and the freezing outside air had flooded into the cabin. The plane continued on autopilot before finally running out of fuel and crashing in South Dakota, but its occupants would have died far earlier from lack of oxygen. It is not the first time such tragedies have occurred and it is unlikely to be the last, for there is simply not enough oxygen at such high altitudes to support life and the failure of a door or window seal may have fatal consequences.

Like Stewart and his colleagues, many of us live life on the edge, often without even realizing it. We routinely fly around the world at altitudes too high to support life, go sailing in frigid waters, expose ourselves to the dangers of the bends when scuba-diving on holiday, or simply live in places where the winter is so severe that it is not possible to survive outside overnight unaided. Environmental extremes are not the prerogative of the adventurous few – with the help of technology, all of us can tolerate severe conditions with equanimity. Without adequate protection, however, it is a very different matter, and every year, thousands of ordinary people die of cold or heat stress, or succumb to mountain sickness.

Yet despite (or perhaps because of) the danger, people have always been fascinated by life at the extremes. Eight hundred million people in fifty-nine different nations watched Neil Armstrong set foot on the moon, and the exploits of polar explorers, mountaineers and other adventurers continue to enthral us. We share vicariously in their dangers, and the more narrowly they cheat death, the greater the thrill. There is even a terrible fascination in tragedy. The poignant story of a climber dying alone high up on a mountain, cut off from help by severe weather, yet still able to use his mobile phone to say goodbye to his wife, touches us more than hundreds killed by floods or earthquakes.

The perils of icy winters, freezing waters and scorching summers were recognized in classical times, but in the late nineteenth and early twentieth centuries the advent of balloons, aeroplanes, submarines and deep-sea diving and the growth in polar and mountain exploration, brought new hazards that required a deeper understanding of human physiology if they were to be circumvented. For many people, like deep-sea divers and astronauts, these risks constitute an unavoidable part of their job. But others put their lives in jeopardy for pleasure. Men – and increasingly women – constantly seek new physical challenges. Our own lives are so cushioned from danger and death that we crave adventure. Rather than a traditional holiday sitting on the beach, many people prefer the adrenaline rush of sports such as off-piste skiing, trekking in the high Andes, scuba-diving, bungee-jumping, and para-gliding. Our ability to tackle these ventures with comparative safety has evolved from a partnership between physiologists interested in how the human body works and intrepid adventurers seeking to push the limits ever further.

This book describes the physiological response of the body to extreme environments and explores the limits to human survival. It considers what happens when you find yourself locked in the freezer, trapped under the ice, or stranded in the desert without water; why an elite mountaineer can climb Everest without supplementary oxygen, yet if an aircraft depressurized at the same altitude its occupants would lose consciousness in seconds; why astronauts may find it difficult to stand without fainting on their return to Earth; why deep-sea divers suffer from bone disease; and other such puzzles. Solving these problems has presented many challenges for physiologists, both physical and intellectual.

The philosopher Heraclitus once remarked that 'War is the father of all things'. As far as the physiology of extreme environments is concerned, he had a point. Military personnel are routinely exposed to adverse conditions – in just the last few years, we have seen wars fought in the freezing Balkan winter, the searing heat of the Kuwaiti desert, and the high mountain passes between India and Pakistan. Many investigations of the effects of heat, cold, pressure and altitude on humans were initiated, directly or indirectly, as a result of this military imperative. It is also salutary to realize that it was not primarily for scientific reasons, but rather because of the Cold War, that humans ventured into space.

Sport – a far more acceptable form of competition between nations than war – has also stimulated much interest in human physiology and in recent years sports physiology has developed into a distinct discipline. Most of us take some form of exercise, even if it is only the occasional sprint for the bus. But there is a limit to how fast we can run, even with training, and exercise imposes its own stresses on the body. This rather different, but related, type of extreme is discussed in Chapter Five.

The scientific study of human physiology is based on controlled experiment. Because the potential hazards may be poorly understood, and the limits to survival unknown, animals are often used in initial experiments to identify the type of dangers involved and to obtain an indication of the safety limits for a person. Ultimately, however, there is no substitute for humans and physiologists have often experimented on themselves – and still do. Some of them even used their children. The eminent scientist J.B.S. Haldane once remarked that his father had used him as a guinea-pig ever since he was four years old (although he did not appear unduly discouraged by this experience, for he followed his father into a distinguished career as a physiologist).

There are good reasons why physiologists use themselves and their colleagues as experimental subjects. It is often easier to understand something by experiencing it yourself than from a second-hand description; and, particularly in the past, the work was often dangerous and unpredictable so that many scientists preferred to take a risk themselves, rather than ask a volunteer to do so. It was also quicker – finding a volunteer takes time. The early physiologists needed considerable courage, as well as skill and scientific curiosity. Sitting in a cramped steel chamber filled with pure oxygen while the pressure is increased, knowing that you are doomed to go into convulsions that may cause you permanent damage, but not knowing exactly when it will happen, is a far from pleasant experience. But as discussed in Chapter Two, such experiments were vital for the safety of deep sea-divers.

People may react very differently to physiological stress, and their behaviour under normal conditions is no indication of the way they will perform under stress: tough commandos can succumb rapidly to mountain sickness, yet their more fragile female companions suffer no ill effects. So while it may not be essential for understanding the scientific principles involved, when it comes to practical applications, the experiments must be repeated on a larger number of volunteers. Unfortunately, not all human guinea-pigs have been volunteers. There are a number of infamous cases in which experiments have been carried out on people without their consent. The Nazis used the inmates of Dachau, the Russians reputedly used prisoners of war, the Japanese experimented on the Manchurian population, and convicted criminals have been used by Western governments even in recent times. Although the latter theoretically may have been volunteers, the choice between execution or reprieve and participation in a, possibly dangerous, experiment, is not really a choice. Moreover, in many cases, the subjects were not fully informed of the risks. Many of these experiments were concerned with testing the effects of chemicals or radiation. But not all. Some were designed to enhance our understanding of how humans cope with extreme conditions. As we shall see, there is also a dark side to the study of life.

Human experiments are still needed, for new types of survival suit for cold water immersion must constantly be tested, and space suits are still a developing technology. But today, experiments are conducted under stringent safety conditions, and the limits to life, obtained from accident and experiment, are well documented.

The study of human physiology has obvious practical applications, but for many scientists (perhaps the majority), the real spur is curiosity; they are driven by Kipling's 'six honest serving men' – by 'What and Where and When, and How and Why and Who'. As a consequence, the life of the physiologist, like that of many experimental scientists, is a curious combination of elation and frustration – elation when a pet hypothesis turns out to be correct, and frustration when, for technical reasons, an experiment does not work and the question it was designed to test cannot be answered. All too often, there seems to be too little of the former and too much of the latter. But piecing together a puzzle, solving an intellectual challenge, or finding a new fact, can be very rewarding, and the sharp excitement of discovery is an exhilaration like no other I have experienced. It is this emotional high that sustains you throughout the long hours needed to obtain the results.

Although many people may find it difficult to appreciate the delights of the scientific life, most will understand the elation of reaching the summit of a mountain, and the sense of achievement that comes from running a marathon. Some physiologists are fortunate because they manage to combine both intellectual and physical adventure. Those seeking to answer questions about how the body works, for example, have often had to go to extreme ends – to the mountain tops, the depths of the sea, the Antarctic icefields, or even into space – to find the answers. The knowledge they gained has been invaluable, for as this book will show, physiology is not just a laboratory science, but something applicable to everyday life. In our battle to survive at the limits a knowledge of physiology, the 'logic of life', is crucial.

Kilimanjaro from the Amboseli Park, in Kenya

Climbing Kilimanjaro

KILIMANJARO IS ONE of the most beautiful mountains in the world. A perfect volcanic cone, it straddles the border between Kenya and Tanzania, rising 5896 metres (19,340 feet) from the African plains. At its feet lies the Amboseli game reserve with its teeming herds of wildebeest, antelope and elephants. Its summit is crowned with ice-fields of breathtaking beauty. Despite its great height, no mountaineering skills are required to reach the top of Kilimanjaro; it is a walk which takes less than three and a half days from base to summit. Unfortunately, the rapidity of this ascent is fraught with danger for the unwary.

We set off through the rain forest early in the morning. The air was warm, heavy and damp, redolent of the tropics. It smelt like the Palm House at Kew. Our feet made little sound on the soft moist earth of the forest floor. Monkeys swung chattering through the canopy far above us. It was difficult to realize that we were climbing all day as we wound our way though the cool dark shade of the forest. Late in the afternoon we emerged from the trees to find a small triangular hut nestling against the side of the mountain in meadows reminiscent of those of the Alps. The sun winked out and night fell almost instantly, since Kilimanjaro lies on the equator.

Next day we climbed to around 3700 metres, crossing high grasslands and passing through vegetation unique to these altitudes in Africa and South America. Giant Senecio, a relative of the common groundsel, towered above our heads. Immense lobelia flowers, like giant blue candles, stood sentinel beside the path. The thinner air was exhilarating, convincing me that I was immune to mountain sickness.

The following morning it was very cold. As we walked, we left the vegetation behind us and entered a high rock saddle hanging between the twin peaks of Kilimanjaro. To our right stood Mawenzi and to our left the higher Uhuru, our ultimate goal. Despite the flatness of the terrain, I felt tired. It seemed a long way across the saddle and even further to the tin huts sited at the foot of the final climb – a giant ash cone.

We spent a third, cold and uncomfortable night at 4600 metres. Sleep was impossible. My head hurt and the world spun around me when I closed my eyes. Despite a lack of appetite, I had forced down lukewarm food and tepid tea (at this altitude, water boils at 80°C), conscious that I would need energy for the coming climb. Now I felt sick. My companions' breath came in jangling gasps interrupted by such long silences that I wanted to shake them awake for fear they had permanently stopped breathing. I waited, shivering, for time to pass.

We rose at two o'clock in the morning to begin the long trek to the summit, for our guide had persuaded us to see dawn break over Mawenzi peak. I now know his real reason for the early start was far more prosaic: we climbed in the dark, so as not to see the enormity of the task that lay before us. The path wound in a shallow zig-zag up a 1200-metre cone of fine grey ash and small stones to the edge of the crater. Even at sea-level, climbing sand dunes is hard work; at this altitude it was torture. For every hard-won three steps up, I slid two steps back. My boots filled with fine abrasive dust. My legs felt unsteady and out of control, so that I staggered wildly, further compromising my progress on the shifting sands. One of my companions collapsed, unable to go further. It is not easy to tell who will succumb to mountain sickness; he was probably the fittest and strongest of our group but now he sat gasping for air like a stranded fish, his only option to descend. We continued, the guide lighting the way ahead with a hurricane lantern held low by his side. Progress was not easy. I fought for breath and struggled to take a few steps between each ever-longer rest. It was only by sheer effort of will and the (quite foolish) determination not to be beaten that I managed the last few hundred feet. I collapsed at the top of the crater rim, my head feeling as if knives were being driven through it, my vision swimming with black dots.

A medley of images danced across my mind. I sat in a dusty Cambridge lecture theatre, shafts of sunlight falling across the desks, listening to a discourse on mountain sickness. What exactly had the lecturer said? It seemed important but it slipped away as brilliantly coloured zig-zags marched majestically before my eyes. The air shivered and a snow leopard slunk around the edge of the ice floes which sail within the crater of Kilimanjaro. It glared at me with yellow eyes and twitched its tail. I looked away and the sun rose, flooding the sky with a soft pink and orange glow, tinting with gold the edges of the thin clouds, Mawenzi peak a sharp black silhouette against a Botticelli sky. I sat on the top of Uhuru's crater rim, the cold wind blowing through my hair, and I knew the illusions were a warning. My brain was slowly shutting down through lack of oxygen. It was past time to leave.

I slithered and slipped drunkenly down the steep slope, suddenly afraid of cerebral oedema, yet fearful of falling forwards and tobogganing uncontrollably downwards if I went too fast. With every step I felt more alive, as oxygen flooded through my brain. I ran the scree, skiing down the mountain in great long slides, slaloming around the rocks and boulders. It took only half an hour to cover the distance I had taken over five hours to climb so painfully.

I was lucky; the previous week two people had died of mountain sickness on the same trek. My own brush with mountain sickness had no permanent effects, but I had been foolish. We had climbed too high too fast: 5896 metres in three and a half days. The high peaks may not be reserved for the gods, but they must be treated with respect.

1

LIFE AT THE TOP

'Great things are done when men and mountains meet;'

WILLIAM BLAKE, *Gnomic Verses*, I

Mount Everest

At 8848 metres (29,029 feet), Mount Everest is the highest mountain on Earth. If it were possible to be transported instantaneously from sea-level to the summit of Everest, you would lose consciousness and lapse into a coma within seconds because of lack of oxygen. Yet in 1978, the Austrian climbers Peter Habeler and Reinhold Messner reached the top of Everest without the aid of supplementary oxygen; and ten years later, more than twenty-five others had also done so. What is the explanation for their apparently impossible feat? The scientific detective story of how the answer to this question was unravelled, the twists and turns along the way, the excitements, extraordinary feats of endurance and colourful characters involved are the subject of this chapter.

Mountains have fascinated and challenged people for centuries. Beautiful but forbidding, they were initially believed to be the home of the gods. The Greek Pantheon lived on the summit of Mount Olympus, the highest mountain in Greece; the Indians considered the Himalayas the abode of the gods; and evidence of ancient human sacrifice, probably to mountain gods, has been found in the Andes. Even today, many cultures hold sacred mountains in reverence – Tenzing Norgay buried chocolate and biscuits on the summit of Everest during the first successful ascent, as a gift to the gods that live there. Mountains lie shrouded in myth and legends, their peaks and crags imaginatively populated not only with gods, but also by mysterious monsters like the Himalayan Yeti and the trauco of southern Chile (that feeds on human blood). Even their names cause enchantment: 'Chimborazo, Cotopaxi, They had stolen my soul away!'[1] Yet despite, or perhaps because of, these stories, people have always been attracted to mountains, whether for spiritual refreshment, the promise of hidden treasure, a means of escaping oppressive regimes, the thrill of exploring new terrain or, more mundanely, to find a way through to the other side: or simply, in George Mallory's memorable phrase, 'Because it's there'.[2]

As a consequence, mountain sickness has been known for centuries. Its cause remained a mystery to the ancients who considered it due to the presence of the gods (which drove men mad), or the result of poisonous emanations from plants, and led to the early European view of mountains as dangerous and mysterious. Some time around the latter half of the nineteenth century, however, mountain climbing emerged as a sport and men vied with the elements and with each other to be the first to reach the highest peaks. Physiologists became increasingly interested in the effects of altitude on the body, and increasingly knowledgeable about their causes, and their studies contributed greatly to the success of the first expedition to reach the summit of Everest. Yet they have been repeatedly astonished by the ability of mountaineers to ascend higher than their predictions.

High altitude is defined, somewhat arbitrarily, as more than 3000 metres (10,000 feet) above sea level. Many people, probably around 15 million, live above this height in the mountainous areas of the world, with the greatest numbers in the Andes, the Himalayas and the Ethiopian Highlands. Many more people visit altitudes of over 3000 metres each year for skiing, backpacking and tourism. The highest permanent human habitations are mining settlements on Mount Aucanquilcha in the Andes, at an altitude of 5340 metres. Although the sulphur mines are located at 5800 metres, the miners prefer to climb the additional 460 metres to work each day rather than sleep higher up. The Indian army is also reputed to have kept troops at 5490 metres for many months, to guard their border with China, but this is probably the limit at which it is possible for humans to live for an extended period, for life at such altitudes is fraught with difficulties. Chief among these is the reduction in the oxygen concentration of the air, but cold, dehydration and the intense solar radiation are also significant problems.

The decrease in the density of the air at altitude means that it contains less oxygen, which poses a considerable problem for most organisms,

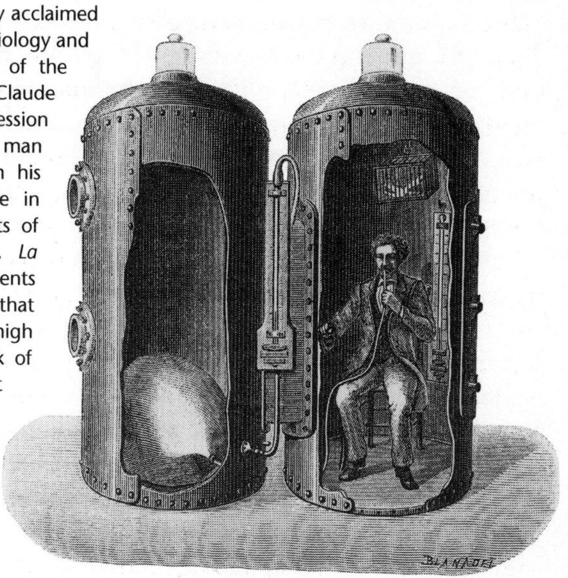

Paul Bert (1833–86) is widely acclaimed as the father of altitude physiology and aviation medicine. A pupil of the famous French physiologist Claude Bernard, he built a decompression chamber large enough for a man to sit comfortably inside in his laboratory at the Sorbonne in Paris, to simulate the effects of altitude. His famous work, *La Pression Barométrique,* presents evidence to support his idea that the deleterious effects of high altitude are due to the lack of oxygen. He was also the first to show that decompression sickness (the bends) is due to the formation of bubbles in the blood (see Chapter Two).

including humans, who need to supply oxygen constantly to all their cells. Within each cell, oxygen is burned, together with foods such as carbohydrates, to produce energy. Cells that do large amounts of work, such as muscle cells, need proportionately more oxygen, and exercise further increases their demands. Oxygen was 'discovered' in 1775, as recounted in Chapter Seven, and its beneficial effects were immediately understood. But it was almost another hundred years before it was recognized, by the Frenchman Paul Bert, that it was a lack of oxygen (hypoxia) that was the main cause of mountain sickness. It took even longer for his idea to become widely accepted.

Early Accounts of Mountain Sickness

The Chinese were the first to document the effects of altitude, in a classic text, the *Ch'ien Han Shu,* that describes the route between China and what is probably Afghanistan around 37–32 BC: 'Again on passing the Great Headache Mountain, the Little Headache Mountain, the Red Land and the Fever Slope, men's bodies become feverish, they lose colour and are attacked with headache and vomiting; the asses and the cattle all being in like condition.' The eminent Chinese scholar Joseph Needham has suggested that such experiences convinced the Chinese that they were meant to stay within the natural borders of their country. Likewise, the Greeks, who found they became breathless on the top of Mount Olympus (around 2900 metres), assumed that the summit was reserved for the gods and was out of bounds to mere mortals.

One of the first clear descriptions of the effect of acute mountain sickness was published in 1590 by Father Jose de Acosta, a Spanish Jesuit missionary who crossed the Andes and spent some time on the high plateau known as the Altiplano. Many of his party became sick when crossing the high pass at Pariacaca (4800 metres). He himself was 'suddenly surprized with so mortall and strange a pang, that I was ready to fall' and considered that 'the aire is there so subtle and delicate, as it is not proportionable with the breathing of man.' He also wrote that at this pass and all along the ridge of the mountains were to be found 'strange intemperatures, yet more in some partes than in others and rather to those which mount from the sea, than from the plaines.' This passage has been taken to indicate that Father Acosta was aware that people who had become acclimatized to high altitude by spending time on the high plains, such as the Altiplano plateau, succumbed less readily to mountain sickness than those who ascended directly from sea-level. Scholars now suggest that this is probably not the case, as the original Spanish text appears to have been incorrectly translated.

The local Inca population, however, were very well aware of the effects of altitude and of how acclimatization took time. They knew that lowlanders died in great numbers if transported to high altitudes to work in the mines and they maintained two armies, one that was kept permanently at high altitude to ensure they were acclimatized, and a second which was used for fighting on the coastal plains. To escape the ravages of the Conquistadores, the Incas retreated higher and higher into the mountains, where the Spanish invaders found it difficult to follow. Although the Spanish eventually established a city at Potosí (4000 metres), it was very much a frontier town and both women and livestock had to return to sea-level to give birth and bring up their off-spring for the first year. The fertility and fecundity of the native women was unaffected but Spanish children born at altitude died at birth or within the first two weeks of life. The first child of Spanish descent to survive was not born until fifty-three years after the city was founded, on Christmas Eve 1598, an event that was hailed as the miracle of St

Nicholas Tolentino. Sadly, none of the 'miracle's' six children survived to maturity. Nevertheless, the problem resolved itself after two to three generations, probably because of interbreeding with the indigenous Indian population. The cattle and horses remained relatively infertile, however, and as a consequence, the Spanish eventually moved the capital to Lima. Infantile mountain sickness is not simply a problem of the past, for it afflicts the lowland Han Chinese colonists of Tibet today.

As the Incas appreciated, mountain sickness is less severe in people who become accustomed to altitude gradually. The dramatic and often fatal consequences of very rapid ascent to high altitude were first encountered by the early balloonists. The first flight in a hot air balloon was made in 1783 by Jean-François Pilâtre de Rozier and the Marquis d'Arlandes in a balloon made by the Montgolfier brothers, Etienne and Joseph. Later the same year another Frenchman, Jacques Charles, invented the hydrogen balloon and reached 1800 metres on his initial ascent, with no apparent ill effects. Balloons are capable of reaching even greater heights, however, which can have serious consequences.

The symptoms of altitude sickness associated with ballooning were described in a famous report by James Glaischer, a meteorologist who accompanied the balloonist Henry Coxwell on a flight from Wolverhampton in 1862. Within an hour they had ascended to a height at which his barometer read 247 millimetres of mercury – around 8850 metres. They continued to rise, but the precise altitude they reached is unclear because above this height Glaischer was no longer able to see the barometer clearly, nor is it certain his barometer was correct; but it is likely to be less than the 11,000 metres he reported. He described vividly how he found his arms and legs were paralysed, he was unable to read his watch or see his companion clearly, he tried to speak but found he could not, and he then became temporarily blind. Finally, he lost consciousness. Fortunately, Coxwell was not completely incapacitated and was able to bring the balloon down, although with great difficulty, by venting hydrogen. Because his arms were paralysed, he had to pull the rope that released the vent valve with his teeth. On the way down, Glaischer recovered consciousness and was able to take notes again at an altitude he calculated as around 8000 metres, which illustrates the rapid recovery that can occur following severe acute hypoxia.

The first fatalities occurred a few years later, in 1875, when three

The famous balloon flight from Wolverhampton of James Glaisher and Henry Coxwell. The lithograph depicts them at the peak of their ascent – an estimated altitude of around 11,000 metres (seven miles high). Glaisher is insensible, collapsed in the basket. Coxwell, who has lost the use of his hands from a combination of hypoxia and cold, is struggling to release the gas valve by pulling the release cord with his teeth. In contrast, the pigeons (in the basket suspended from the ring) seem unaffected by the altitude.

Lithograph of H.T. Sivel, Gaston Tissandier and J.E. Croce-Spinelli in the balloon *Zenith*. Sivel (left) is about to cut the strings holding the bags of ballast in order to ascend higher. Tissandier (centre) is reading the barometer. Croce-Spinelli has the mouthpiece of the oxygen equipment in his hands; this is connected to the striped balloon, which contains a mixture of 72 per cent oxygen in air.

The balloon took off on 15 April 1875 from a gas factory at Villette, outside Paris, and ascended to a height of 7500 metres. At this point, pictured right, Sivel asked his companions if they should go higher and, on receiving their assent, released the ballast. The balloon then rose rapidly to 8600 metres. All three men became paralysed and passed out before they felt the need to breathe oxygen. Both Tissandier and Croce-Spinelli briefly regained consciousness at separate times but, confused by hypoxia, they each let go more ballast, which only exacerbated their situation for it caused the balloon to rise further. When Tissandier finally awoke, the balloon was at 6000 metres and falling rapidly, but both his companions were dead.

French scientists, Sivel, Tissandier and Croce-Spinelli, ascended to over 8000 metres in the balloon *Zenith*. Although they had primitive oxygen equipment, the amount of oxygen they carried was small and they agreed not to use it until they felt it was really necessary.[3] Unfortunately, the over-confidence and feeling of well-being characteristic of acute oxygen starvation meant they never used it and they all lost consciousness. Only Tissandier survived. He later related that he tried to use the oxygen equipment but was unable to move his arms. However, far from feeling concerned, he wrote: 'one does not suffer in any way; on the contrary. One feels an inner joy, as if filled with a radiant flood of light. One becomes indifferent and thinks no more of the perilous situation or of the danger.'

The Ascent of Everest

With the advent of mountaineering, the effects of mountain sickness became more widely known and better understood. By the mid-1920s it was appreciated that people could climb as high as 8000 metres and remain there safely for a few days, providing they had spent many weeks at an intermediate altitude gradually acclimatizing. In contrast, when exposed to a similar barometric pressure in a decompression chamber, consciousness was lost within a few minutes.

The 1953 British expedition to Mount Everest, led by Sir John (later Lord) Hunt, was well aware of the importance of acclimatization. The long march from Kathmandu to Khumbu, at the foot of the mountain, took several weeks and imposed an obligatory period of acclimatization because most of the trek is at 1800 metres, rising only occasionally to 3600 metres. A further four weeks was then devoted to acclimatization in the Khumbu district (4000 metres) before attempting to establish camps higher up the mountain. The team also adopted a policy of siting these camps at altitudes at which it was possible to sleep and eat easily, and of going down to lower altitudes for rest periods of a few days to recover, a procedure that is copied by most modern expeditions and, as we shall see, has a sound physiological basis.

For the first time, there was also a comprehensive policy on the use of supplementary oxygen; previously, oxygen was not widely used because most climbers had little confidence in the new-fangled gear and the early equipment was very heavy. Above 6500 metres, the Everest expedition used oxygen, both to assist in sleeping (at a rate of 1 litre per minute) and when climbing (4 litres/minute). Even with this advantage, the effects of altitude caused a gradual physical deterioration and they all lost weight. Sometimes they became severely incapacitated, as graphically described by Hunt:

> 'Our progress grew slower, more exhausting. Each step was a labour, requiring an effort of will to make. After several steps at a funereal pace, a pause was necessary to regain enough strength to continue. I was already beginning to gasp and fight for breath . . . My lungs seemed about to burst; I was groaning and fighting to

get enough air; a grim and ghastly experience in which I had no power of self-control.'

The cause of this extreme difficulty was discovered later. The tube connecting Hunt's face-mask to the oxygen bottles was completely blocked with ice so that he received no oxygen; not only was he carrying the heavy oxygen equipment but he was gaining no benefit from it! In his account of the Everest expedition, Hunt later wrote: 'I would single out oxygen for special mention . . . only this, in my opinion, was vital for success. But for oxygen, we should certainly have not got to the top.'

News of the conquest of Everest on 29 May 1953 by Edmund Hillary and Sherpa Tenzing Norgay arrived in London on 2 June, just in time

Tenzing Norgay photographed on the summit of Everest by Edmund Hillary on 29 May 1953 – the first successful ascent.

for the Coronation of Her Majesty Queen Elizabeth. It was announced over the loudspeakers along the Coronation route and greeted by wild cheering of the crowds. At Base Camp, the successful party were amazed to hear the news of their achievement announced on the All India Radio, as the reporter, James Morris of *The Times,* had only left Advance Base Camp on 30 May to file his article. To celebrate, they fired twelve mortar bombs, a gift of the Indian army, into the snows.

The use of oxygen to conquer Everest led to the belief that it was not possible to survive on its summit unassisted. Indeed, Dr Griffith Pugh, a physiologist on the first expedition to reach the summit of Everest, claimed that 'only exceptional men can go above 8200 metres without supplementary oxygen'. His statement was supported by a number of tragic accidents in which elite mountaineers, climbing unassisted, died, usually from exhaustion brought on by hypoxia that caused them to stagger wildly and slip to their deaths. As has so often been the case with high altitude physiology, however, the resilience and determination of the mountaineers proved the scientists wrong, for in 1978 Peter Habeler and Reinhold Messner climbed Everest without oxygen. Since then their remarkable achievement has been repeated by many others, including, in 1988, the first woman, Lydia Bradey (her claim is disputed, since she climbed alone and thus it could not be proved she actually made it to the top).

It is clear from these accounts that a distinction must be made between the physiological effects of sudden ascent to altitude, such as might occur in a balloon flight or when the cabin pressure of an aircraft is suddenly lost, and the effects of a more gradual ascent, typified by the slow climb to the summit of a mountain, in which time is taken for acclimatization. The effects of life-long residence at high altitude constitute yet a third case.

A Digression on Barometric Pressure

Evangelista Torricelli was the first to conceive that air has weight. In a letter to a colleague dated 1644 he wrote that 'we live submerged at the bottom of an ocean of the element air, which by unquestioned experiments is known to have weight'. A pupil of Galileo, Torricelli is also credited with making the first mercury barometer for measuring atmospheric pressure (the pressure exerted by the weight of the air itself).

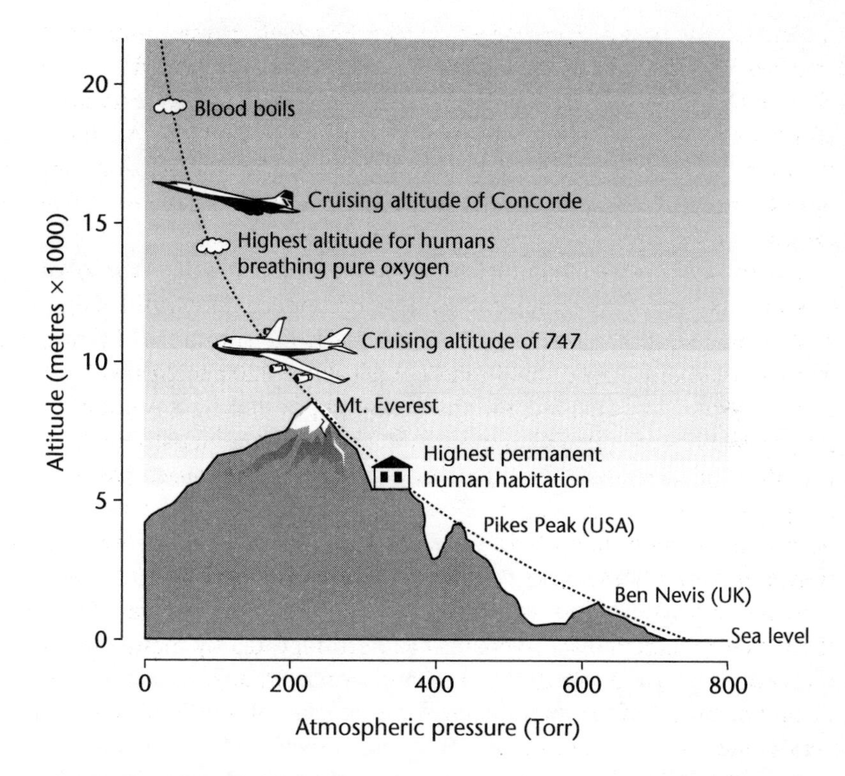

The effect of altitude on atmospheric pressure and on the partial pressure of oxygen in the air. The fall in atmospheric pressure with altitude is not linear because air is compressible and is squashed by the weight of air on top of it. Thus the pressure increases more rapidly close to the ground.

The decrease in the density of the air with altitude means that the atmospheric pressure decreases as one ascends higher and higher. This was first shown in 1648 by Blaise Pascal on the Puy de Dome in what he engagingly described as 'The Great Experiment'. Simply put, the pressure is less the higher one goes because there is less weight of air pressing down on top.

Until very recently, the units used to measure atmospheric pressure (Torr) were named after the Italian Torricelli in recognition of his important contribution. Officially, Torr have now been superseded by a new unit of pressure named after the Frenchman Pascal – a change that, as might be imagined, was not without controversy. However, because

Blaise Pascal (1623–62) is credited with being the first person to demonstrate scientifically that the atmospheric pressure falls with altitude. Pascal did not actually carry out the experiment himself, but instead persuaded his brother-in-law and various local dignitaries to climb the Puy de Dôme in central France carrying a barometer, and measure whether the pressure fell. Another instrument was left in the town of Clermont under the guard of the Reverend Father Chastin (memorably reported to be 'as pious a man as he is capable'), as a control. Only the barometer that was taken to the summit was observed to change.

much of the early literature uses Torr, and many physiologists continue to do so, I have also done so here.

At sea-level, the atmospheric (or barometric) pressure is around 760 Torr (millimetres of mercury). Oxygen makes up 21 per cent of the air, 0.04 per cent is carbon dioxide,[4] and the rest is mostly nitrogen. Thus at sea level the pressure produced by oxygen, known as the partial pressure of oxygen, is 159 Torr (21 per cent of 760 Torr). On the summit of Everest the air contains the same percentage of oxygen, but because the barometric pressure falls to about 250 Torr, the partial pressure of oxygen in the air is reduced proportionately. Moreover, the relative decrease in the partial pressure of oxygen in the lungs is even greater than that in the atmosphere. This rather surprising fact arises because the body produces a significant amount of water vapour. Its presence in the alveoli – the small air-sacs where gas exchange between the air in the lungs and that dissolved in the blood takes place – limits the space available for oxygen, a fact that becomes increasingly important at altitude.

At any altitude, the air in the lungs is saturated with water vapour produced by the body. This can be seen very clearly on a cold day, when the water vapour you breathe out condenses in the cold air, forming a small cloud. Water vapour has a partial pressure of 47 Torr. This means that when the atmospheric pressure is 47 Torr, which occurs at an altitude of 19,200 metres, the lungs will be entirely occupied by water vapour, leaving no room for oxygen or other gases. The fraction of the gas pressure in the lungs that is due to water vapour thus increases with altitude, rising from 6 per cent at sea-level to 19 per cent on the summit of Everest.

The presence of water vapour in the alveoli helps explain why the partial pressure of oxygen is lower in the air-sacs than in the atmosphere (the fact that oxygen is extracted by the body is also a contributing factor). It also physically limits the altitude to which humans can ascend, even when breathing pure oxygen. The lowest barometric pressure at which the normal oxygen concentration of the lungs (100 Torr) can be maintained when breathing pure oxygen is about 10,400 metres, which is around the cruising altitude of most commercial aircraft. It is possible to survive at higher altitudes because an increase in breathing blows off some of the carbon dioxide in the lungs and so provides more room for oxygen. Above 12,200–13,700 metres, however, insufficient oxygen can be provided and consciousness is lost. Above 18,900 metres, the blood 'boils' (it actually vaporizes) at body temperature. This explains why a pressurized suit or cabin, with a self-contained air supply, is required for very high altitude or space exploration (see Chapter Six).

The Perils of Sudden Depressurization

'In the event of a sudden loss of cabin pressure, oxygen masks will fall from the overhead lockers'. The enormous increase in air travel during the past 25 years means that most of us are familiar with these words, although fortunately few of us have experienced such an emergency. Most commercial aircraft cruise at a height of around 10,400 metres. If a window blows out at this altitude, it will be accompanied by a loud bang as the air rushes out of the cabin and the pressure equilibrates with that of the outside air. Loose objects, and people who are not strapped in, may be sucked out and the cabin fills with a fine mist as the temperature drops to that of the outside and the water vapour in the air condenses. It is vital to put on your oxygen mask quickly as the oxygen level in your lungs will fall precipitously and you will lose consciousness in under thirty seconds. 'Useful' time, during which the pilot is capable of taking corrective action, is even less – only about fifteen seconds. One airline pilot passed out because he dropped his spectacles when the cockpit suddenly depressurized and he bent down to pick them up before putting on his oxygen mask. Luckily, his co-pilot did not make the same mistake.

The partial pressure of oxygen in the lungs at 10,400 metres when breathing unpressurized air is about 20 Torr, too low to support life. When breathing pure oxygen, however, it rises to around 95 Torr. This is sufficient for survival if you are sitting quietly but not if you exert yourself, which is one reason why cabin staff are trained to sit down until the plane has levelled out at a reasonable altitude (the other reason is that the plane is put into a steep dive in order to lose height rapidly).

The low exercise capacity at altitude was illustrated rather dramatically during the early part of the Second World War. Although the rear-gunners of bomber aircraft flying at 5500 metres were quite alert when sitting in their gun turrets breathing air, when they attempted to crawl back into the main part of the aircraft many of them passed out. This was because the increased demand for oxygen by the working muscles could not be matched by increased uptake from the air, and the oxygen available for the brain fell below that required for consciousness. Providing one is sitting quietly, however, it is possible to ascend to 7000 metres while breathing air in an unpressurized aircraft before losing consciousness; an altitude that, it should be noted, is significantly lower than the summit of Everest.

More insidious than an instantaneous depressurization is a slow loss of cabin pressure, because the progressive reduction in oxygen concentration may not be readily apparent. The pilot may be unaware that anything is going wrong and so fail to take corrective action. As the early balloonists so vividly described, such gradual oxygen starvation can generate feelings of euphoria, and lead to loss of concentration and impaired judgement. Eventually, it causes reduced muscle power, unconsciousness, coma and death. These effects result from the inability of the body to adjust sufficiently rapidly to the lowered oxygen concentration of the air at altitude.

The legal limit for flight without oxygen in an unpressurized cabin is 3000 metres, although oxygen is usually used above 2400 metres to ensure a good safety margin. Commercial aircraft are pressurized to an altitude of 1500–2400 metres rather than sea-level, because the weight and cost needed to maintain a greater pressure difference across the cabin walls is prohibitive. It is also unnecessary, because at this altitude the partial pressure of oxygen in the air is sufficient to ensure that the blood will normally be completely saturated with oxygen. Patients suffering from heart or lung diseases, however, may not be able to cope with the reduced oxygen levels and may need supplementary oxygen in flight. Adjustment of the cabin pressure to that on the ground, and vice versa, is the reason for the 'popping' of the ears that passengers experience during landing and take-off from sea-level (this phenomenon is explained in more detail in Chapter Two).

Unlike commercial aircraft, many high-performance fighter planes are not pressurized, or only pressurized to an altitude of 7600 metres, because the excess weight produced by fully pressurizing the cockpit would make them much less manoeuvrable. As a consequence, the pilot must wear a tight-fitting face-mask and breathe a mixture of air and pure oxygen. The mixture is automatically adjusted according to altitude to ensure that the pilot receives sufficient oxygen but not so much as to cause oxygen toxicity (see Chapter Two). Above 11,500 metres they must be supplied with pure oxygen under pressure. Breathing pressurized air feels strange: in contrast to normal breathing, where inspiration is an active process and expiration occurs when the chest muscles relax, pressurized air fills the lungs passively and must be actively forced out. Thus, pressure breathing can be quite hard work. A further problem is that the lungs may burst if the gas pressure is increased too high, rather like the self-important frog in Aesop's fable who puffed out his chest until he popped. If an external counter-pressure is provided to support the chest wall, however, the lungs can tolerate higher gas pressures and air-force pilots therefore wear a counter-pressure suit at altitude. This is basically a tight-fitting garment that inflates with air around the chest and abdomen at low atmospheric pressure. It is used by military pilots above 12,000 metres because of the danger of explosive decompression if the canopy of the aircraft is cracked (by shrapnel, for example). A similar suit was worn by Judy Leden when she launched her hang-glider from a balloon 12,000 metres above the Jordanian desert in 1996, and in so doing broke the world altitude record for a hang-glider.

Civilian aircraft are designed so that if a window fails it takes many seconds for the air to escape and the pressure to fall (this is one reason why the windows of Concorde are so small). If a combat aircraft is hit by a missile, however, or its pilot is forced to make an emergency exit by ejecting through the canopy at altitude, decompression can take place very rapidly. Pilots are therefore trained to breathe out throughout decompression, to prevent the attendant expansion of air rupturing their lungs. They are also at risk of the 'bends', which results when gases dissolved in the body fluids bubble out of solution at low pressure. The problems of gas expansion on decompression at altitude resemble those experienced by divers ascending from the depths, and are considered more fully in Chapter Two.

In contrast to most other commercial aircraft, Concorde cruises at a height of 15–18,000 metres. Even when breathing pure oxygen under pressure, this is well above the limit at which it is possible to survive (the ceiling is around 14,000 metres). As explained earlier, the low barometric pressure at these altitudes means there is simply not enough room in the lungs for the necessary amount of oxygen. It is also close to the limit at which the body fluids start to vaporize at body temperature (18,900 metres). A sudden depressurization at Concorde's cruising altitude is therefore likely to be fatal, a fact of which many passengers may be blissfully unaware.

Acute Mountain Sickness

Although few people are likely to have experienced an aircraft depressurization, the increased ease of travel and the popularity of adventure holidays in recent years mean that many are now familiar with the effects of mountain sickness. The journey to the foot of Everest has become a routine tourist trek, thousands of inexperienced individuals have made it to Base Camp, and a marathon is regularly run down the flanks of the mountain. In the Andes, large numbers of people each year take the Inca trail from Cusco to the ancient city of Machu Pichu, that winds through spectacular passes as high as 4500 metres. Because it is possible to reach the High Andes directly by rail, or plane, mountain sickness is common. Air travellers to La Paz, the capital of Bolivia, which is located at 3500 metres, are advised not to exert themselves too much on arrival, but several businessmen die each year from heart attacks or thromboses precipitated by the high altitude.

Symptoms of mountain sickness usually occur in lowlanders on ascent above 3000 metres but, given time, most people are able to adjust. Above 4800–6000 metres, the height of the highest Himalayan and Andean communities, however, further acclimatization becomes impossible and all aspects of performance gradually deteriorate. Even for the most acclimatized individuals, ascent above 7900 metres is dangerous and must be limited to a few hours. Climbers refer to this altitude as the Death Zone because a prolonged stay causes rapid physical deterioration. This is the reason expeditions camp at lower altitudes and mount a final dash to the top, aiming to stay above 7900 metres for as short a time as possible.

Mountain sickness begins within eight to forty-eight hours after rapid ascent to high altitude. Initially, you feel light-headed, often euphoric, as if drunk on the rarefied air. After a few hours, however, this wears off and you feel unaccountably tired; it requires an unusual effort to walk and running is simply not an option. The difficulty in walking is exacerbated by feelings of dizziness which may cause you to lose your balance. It is hard to sleep and you wake up abruptly many times throughout the night, often with the unpleasant feeling of being suffocated. You have a severe headache, lose your appetite, feel nauseated, and may even vomit. Haemorrhage of the small blood vessels in the retina of the eye is common, but these usually heal leaving no permanent damage.

In most people, these unpleasant symptoms disappear after a few days. On occasion, however, they may progress to life-threatening pulmonary oedema, in which the lungs become filled with fluid. More rarely, the brain swells, a condition known as cerebral oedema in which the victim complains of a splitting headache, loss of balance and an intense desire to lie down and do nothing; this is rapidly followed by coma and death. Although oxygen can be beneficial in mountain sickness, and in pulmonary and cerebral oedema, the only real cure is rapid descent to lower altitudes. To pay a porter to carry you further up the mountain, as some tourists in the Himalayas have been known to do, is a fatal mistake.

A graphic first-hand account of the disabling effects of mountain sickness has been given by Edward Whymper. During the first ascent of Chimborazo in 1879, he and his guides, Jean-Antoine and Louis Carrel, were incapacitated by the rarefied air at around 5000 metres.

> 'In about an hour I found myself lying on my back, along with both the Carrels, placed *hors de combat,* and incapable of making the least exertion. We knew that the enemy was upon us and that we were experiencing our first attack of mountain sickness. We were feverish, had intense headaches and were unable to satisfy our desire for air, except by breathing with open mouths. This

naturally parched the throat. . . Besides having our normal rate of breathing largely accelerated we found it impossible to sustain life without every now and then giving spasmodic gulps, just like fishes when taken out of water.'

Around 40 per cent of people who go trekking above 4000 metres will experience some degree of mountain sickness, although not all as severely as Whymper and the Carrels. It is not easy to predict who will succumb, for it is unrelated to fitness – elite paratroopers can be incapacitated while their frail grandmothers are unaffected. The cause of acute mountain sickness is not certain, but both the low oxygen

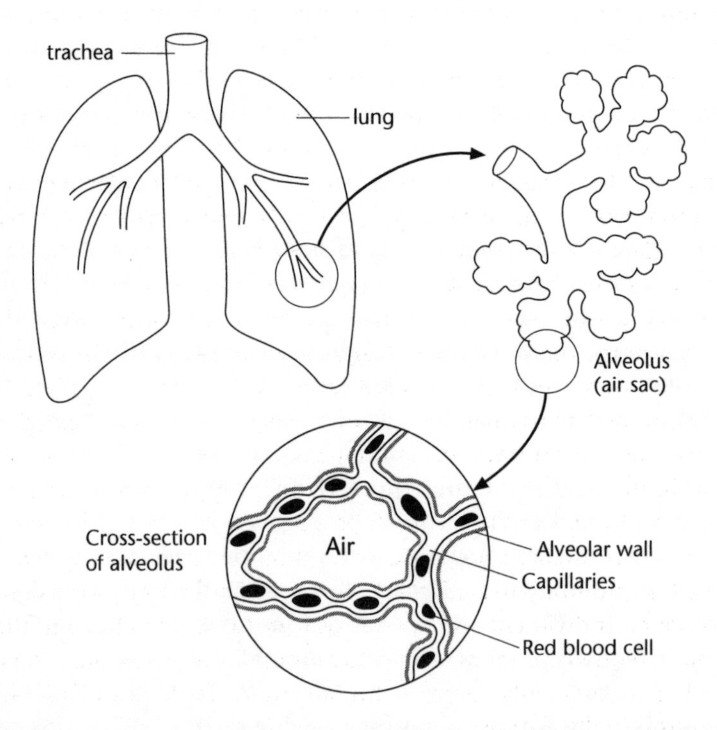

The lungs consist of a series of branching pipes that become finer and finer with every division and end in small air-sacs, called alveoli. There are about 150 million alveoli in each lung. Their walls are very thin and are surrounded by a network of the finest blood vessels (called capillaries), so that the flow of blood in the alveolar walls has been likened to a 'sheet' of flowing blood. It is at this interface that gas exchange takes place between the air in the alveoli and the blood in the capillaries. The surface area of the alveoli is immense, almost 70 square metres or about the size of a tennis court.

concentration in the blood and the reduction in the acidity of the blood (see below) are thought to be important. Some investigators believe that these lead to shifts in the body fluids and to mild cerebral oedema. Measurements of cerebral blood flow, which have been made at altitudes as high as 5300 metres, support this view.

Pulmonary oedema, when the lungs fill with fluid, seems to result from the reaction of the pulmonary blood vessels to the low oxygen levels in the lung that occur at altitude. At sea-level, a low oxygen concentration in a single air-sac (an alveolus) usually signifies that the air flow is obstructed. Since it is clearly inefficient to perfuse this alveolus, the local blood vessels constrict, shutting off the flow of blood and diverting it to other regions that are better ventilated. Unfortunately, they cannot distinguish between a low alveolar oxygen concentration due to a reduced air flow and one resulting from a decrease in the partial pressure of oxygen in the inspired air. Inevitably, therefore, the pulmonary blood vessels constrict at altitude. However, some vessels are more sensitive to low oxygen than others, so vasoconstriction is only patchy. This forces more blood through the capillaries that remain open, producing an increase in pulmonary blood pressure that causes fluid to leak out of them and accumulate in or between the air-sacs. The situation is similar to that which occurs when some of the holes in a showerhead get blocked by limescale – the pressure of the water that comes out of the ones that are not blocked is much higher. Because no fluid leaks from the hypersensitive capillaries (which are of course closed), the oedema is patchy – as one expert memorably remarked: 'the lung looks as if it is full of cannonballs'.

Fluid in the air-sacs interferes with gas exchange. Breathing becomes laboured and sharp crackles can be heard at the bottom of the lungs, possibly caused by the squelching of fluid in the lungs during respiration. Unless promptly treated, the victim in effect drowns in the liquid they produce. Individuals who ascend quickly to 3000 metres and then carry out strenuous physical activity are especially prone to pulmonary oedema: it occurs only rarely if the ascent is gradual and physical exertion is initially avoided.

Of considerable importance to climbers, and to those who live and work permanently at altitude, is their capacity to do work. Clearly, the harder you are working (the faster you climb), the more oxygen will be needed. For lowlanders, the ability to do work falls rapidly at altitude: the work capacity at 7000 metres is less than 40 per cent of that at sea-level. Without oxygen, the rate of climb can be very slow indeed: in 1952 it took Raymond Lambert and Tenzing Norgay five and a half hours to climb just 200 metres on the South Col of Everest; and at the summit of the mountain, Reinhold Messner found that he and Peter Habeler collapsed into the snow with exhaustion every few steps, so that the last 100 metres took over an hour.

> 'After every few steps, we huddle over our ice axes mouths agape, struggling for sufficient breath to keep our muscles going. . . . at a height of 8800 meters, we can no longer keep on our feet while we rest. We crumple to our knees, clutching our axes. . . . Every ten or fifteen steps we collapse into the snow to rest, then crawl on again.'

Similar difficulties are experienced by unacclimatized individuals at lower altitudes. Permanent residents, however, have a remarkable work capacity. Air travellers arriving in La Paz feel instantly exhausted due to the thin air and are surprised (and chagrined) to discover the locals running a marathon!

Into Thin Air

The first thing you notice when you arrive at high altitude is that you breathe faster. This increase in respiration[5] is an immediate and important response to the reduction in the partial pressure of oxygen in the air, and it enables more oxygen to be delivered to the tissues. It is brought about by chemoreceptors (the carotid bodies) located in the carotid arteries, which sense the reduced level of oxygen in the blood and signal the respiratory centre in the brain to increase breathing. The carotid bodies are sited at an important position, because they monitor the oxygen concentration of the blood entering the brain.[6] The mechanism by which they sense the change in oxygen is still hotly debated.

The initial increase in respiration is never very large – no more than 1.65 times that at sea-level, even at heights as great as 6000 metres. This is because hyperventilation of the lungs not only enhances oxygen uptake, it also causes more carbon dioxide to be lost during expiration. Carbon dioxide is produced by the body as a waste product of metabolism, in very considerable amounts. It dissolves in solution to give carbonic acid and the quantity of the gas expired is equivalent to 12.5 litres of industrial strength acid (or, more correctly, 12.5 moles of hydrogen ions) each day! Carbon dioxide is transported by the blood from its site of manufacture in the tissues to the lungs, where it is blown off into the air. Its concentration in the air-sacs therefore varies with the rate of breathing: higher respiratory rates will blow off more carbon dioxide and reduce both the alveolar and the blood concentrations of the gas.

Carbon dioxide is a powerful regulator of breathing (it acts on a different set of chemoreceptors, found in the brain) and if its concentration in the blood falls, breathing is inhibited. It is possible to demonstrate this for yourself. You will find you are able to hold your breath for longer if you breathe very rapidly for a short time beforehand. (Do not do this for more than a minute or you may become dizzy.) The reason is that breath-holding is terminated not by the demand for oxygen but rather by the rising carbon dioxide concentration in the blood. When this reaches a critical level it stimulates you to take a breath. Hyperventilation before holding your breath blows off carbon dioxide from the body and enables a longer period to elapse before carbon dioxide builds up to a level sufficient to stimulate breathing. The antagonistic drives from oxygen and carbon dioxide explain why no change in breathing occurs at altitudes of less than 3000 metres.

The switch from oxygen control to carbon dioxide control of breathing is not always smooth and can result in 'hunting' or 'oscillations' analogous to those seen in poorly-adjusted central heating systems. This manifests as alternate periods of breathing and breath-holding that can be disturbing – and distressing to one's companions. It occurs more commonly at night. The explanation for this peculiar pattern of breathing is that the increased rate of breathing occasioned by the low oxygen concentration of the air results in the loss of carbon dioxide from the body and thus breathing stops. There follows a variable length of time during which carbon dioxide builds up again in the blood, so alleviating this inhibition, while at the same time the oxygen demand becomes increasingly strong. Breath-holding is terminated by a sudden gasp for air, which is sometimes sufficiently violent to wake the sleeper, and the cycle then repeats itself. The constant interruptions to sleep contribute to the difficulties of life at altitude and explain the climbers' maxim, 'Climb high, sleep low'.

The reduction in the carbon dioxide concentration in the blood that results from increased breathing has the effect of reducing the blood concentration of hydrogen ions (also referred to as a reduction in the acidity of the blood, an increase in blood pH, or an increase in blood alkalinity). This is because carbon dioxide combines with water to produce bicarbonate and hydrogen ions, in a reaction catalysed by an enzyme called carbonic anhydrase. It is thought that it is the hydrogen ions produced by this reaction that actually regulate the rate of breathing, rather than carbon dioxide itself. The chemoreceptors that sense the change in hydrogen ion concentration are located at the base of the brain, in a region known as the medulla.

Flying High

At the summit of Everest a person can only survive without oxygen if they are very fit and have taken time to adapt. Even then, they move slowly and with difficulty. In contrast, birds such as the bar-headed goose (*Anser indicus*) regularly migrate across the Himalayas flying at the same or higher altitudes. Moreover, they may begin their flight at sea-level and reach altitudes of 9000 metres in less than a day, so that they have no time to acclimatize. Even a common house sparrow is alert and active when pressurized to 6000 metres, while a human, by contrast, would be comatose. So what underlies the extraordinary capacity of birds to tolerate low oxygen levels?

One reason seems to be that the bird lung is designed differently from that of humans and is able to extract more oxygen from the inspired air and deliver more carbon dioxide to it. The lungs of a bird are small and compact, but they communicate with extensive air spaces that extend between the internal organs and into the bones of the skull and skeleton. These air spaces do not act as respiratory surfaces but rather as storage sacs. The fine tubes which connect the posterior and anterior air spaces are the places where gas exchange takes place (i.e. these are the lungs).

It takes two full breaths for air to pass completely through the lungs of a bird. Inspiration first fills the posterior air-sacs. During expiration and the next inspiration, this air then passes to the anterior air-sacs and oxygen is extracted *en passant*, as it moves through the lungs. Finally, the air is expelled from the anterior air-sac on the next breath out. This adaptation means that the air flows continuously over the respiratory surfaces, enabling the bird to extract far more oxygen than a mammal. In the latter, the blindly-ending alveoli mean that air is not forced across the

Why should breathing in humans be regulated principally by carbon dioxide rather than by oxygen? The reason seems to be that we evolved at sea-level and, from an evolutionary viewpoint, only ventured up high mountains very recently. At sea-level, the oxygen concentration in the lungs is far higher than is needed, even if breathing were to decrease substantially. On the other hand, the rate of breathing has a very pronounced effect on the carbon dioxide concentration in the lungs and tissues, and it is therefore very important to match the respiratory rate to the concentration of the gas in the body. Hence, carbon dioxide acts as the main controller of breathing.

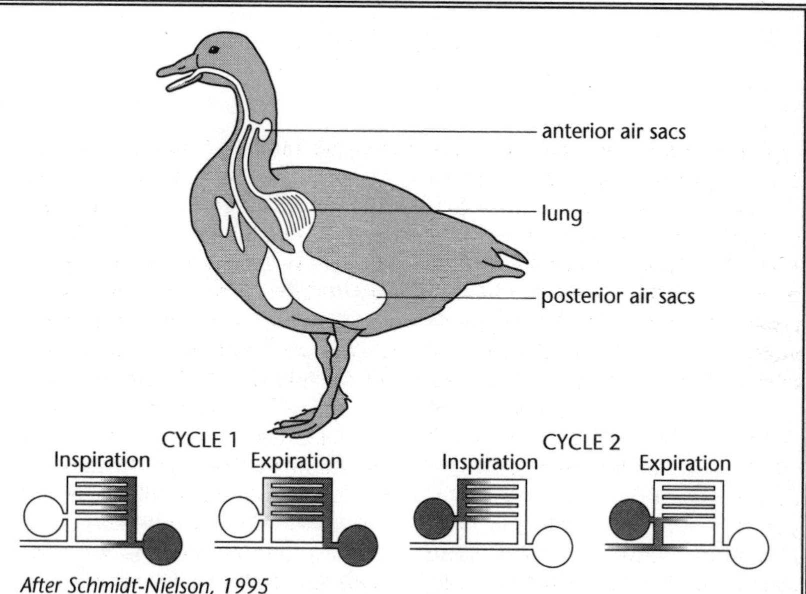

After Schmidt-Nielson, 1995

gas-exchange surface but instead must slowly diffuse to it.

Another factor that helps birds fly at altitude is that they are much less sensitive to the fall in the blood carbon dioxide concentration and the accompanying reduction in blood acidity than are mammals. Thus they maintain an elevated respiratory rate even when blood carbon dioxide levels fall. Birds also have larger hearts, which pump more blood per beat than a mammal of a comparable size; and the haemoglobin of birds that live at high altitudes binds oxygen more avidly, so that more oxygen is extracted from the air.

Acclimatization

Although the increase in breathing on arrival at high altitude is comparatively modest, over the next week or so it increases further, finally becoming as high as five to seven times normal after two to three weeks. This secondary rise in respiration is the most important adaptation to altitude, and it determines how high an individual will be able to climb; the faster and deeper you breathe, the more oxygen is taken up, and the higher up the mountain you can go.

Acclimatization entails removal of the brake on breathing initially imposed by the reduced level of carbon dioxide in the blood and the

Haemoglobin

Haemoglobin is a globular molecule that is made up of four subunits. Each of these subunits is in turn composed of a haeme moiety bound to a globin polypeptide. At the heart of the haeme ring sits an iron atom, to which oxygen attaches. It is haeme that is responsible for the colour of blood. When bound to oxygen (oxy-haemoglobin), haemoglobin is a brilliant scarlet which accounts for the colour of arterial blood and also for the pink colour of 'white' humans, who have translucent skin. Deoxy-haemoglobin is the dark purplish-blue colour characteristic of venous blood. This colour is also known as cyan – hence cyanosis, the technical term for the bluish colour of the lips and extremities of individuals who suffer from poorly oxygenated blood. The brown colour of dried blood or old meat is that of met-haemoglobin. This is oxidized haemoglobin (as opposed to oxy-haemoglobin). It occurs when the iron atom at the heart of the haemoglobin molecule is oxidized from its normal ferrous (Fe^{2+}) form to a ferric (Fe^{3+}) ion, which does not bind oxygen. Red blood cells possess an enzyme that converts the small amount of met-haemoglobin which forms spontaneously back to the normal type of haemoglobin. A bright cherry-red colour to the blood signifies carbon monoxide poisoning, a condition in which a carbon monoxide molecule usurps the place at the centre of haemglobin normally reserved for oxygen. Poorly-adjusted gas appliances that generate carbon monoxide may strikingly lower, or even abolish, the oxygen-carrying capacity of the blood. In these accompanying decrease in blood acidity. Clearly, restoration of the blood acidity would be beneficial in this respect, and in fact this is brought about by the kidneys.[7] But although such renal compensation is undoubtedly important for long-term acclimatization, it cannot be the whole story, for the rate at which it occurs is too slow, and its effect is too small, to account for the increase in respiration observed during the first few days at high altitude. An additional process, as yet unidentified, must therefore be involved (both an increased sensitivity of the carotid bodies to low oxygen and a gradual restoration of the acidity of the fluid that surrounds the central chemoreceptors in the brain have been postulated).[8] Given its importance, it may be surprising to learn that the mechanism responsible for the secondary rise in respiration is still not clearly resolved – but it provides physiologists with an excellent

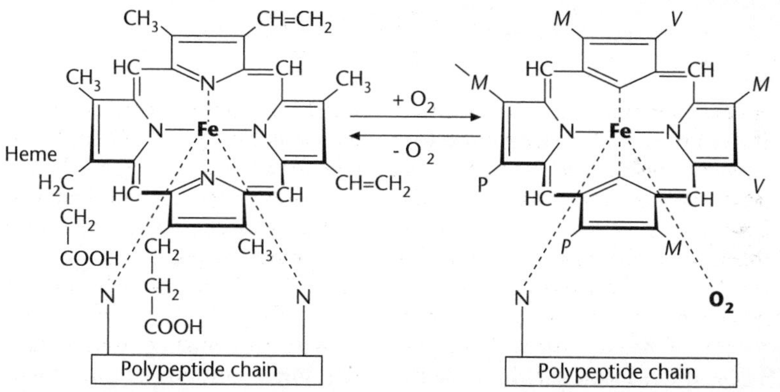

Deoxgenated hemoglobin

Oxyhemoglobin

circumstances, the only remedy is to give the patient pure oxygen to breathe. Better still is to also put them in a hyperbaric chamber, for at 3 atmospheres pressure sufficient oxygen dissolves in the blood to maintain life until carbon monoxide has released its hold on the haemoglobin molecule. Because oxygen is a serious fire risk, the chamber is filled with air and the patient is given oxygen by mask.

Haemoglobin is a famous molecule, with a string of 'firsts' to its name. It was one of the first proteins to be crystallized, to have its molecular weight accurately determined, and to be shown to have a specific physiological function (oxygen transport). It was also the first protein for which the three-dimensional structure was determined, by X-ray analysis of the haemoglobin crystal by Max Perutz in 1959.

excuse for journeying to the tops of mountains to try to sort it out.

Hyperventilation is the key to why it is possible for an acclimatized mountaineer to survive on the top of Mount Everest without supplementary oxygen. As Reinhold Messner memorably put it, he was 'nothing more than a single gasping lung' when he reached the summit. When you breathe faster, more carbon dioxide is blown off, which lowers the partial pressure of carbon dioxide in the lungs and provides more space for oxygen. It turns out that as elite mountaineers ascend ever higher and higher the partial pressure of carbon dioxide in their lungs falls dramatically, until on the summit of Everest it is only 10 Torr instead of the 40 Torr found at sea-level. Not everyone can acclimatize sufficiently to generate the enormous increase in respiration necessary to reduce their carbon dioxide level this low, nor can they tolerate the fall in the blood acidity that accompanies it. These people will never make it to the top, as their failure to blow off sufficient carbon dioxide means that they do not have enough space in their lungs for oxygen. Even for those who are successful, a considerable acclimatization period is required before their bodies are able to tolerate these very low carbon dioxide levels.

The partial pressure of *oxygen* in the lungs of a well-acclimatized climber standing on the summit of Everest is around 36 Torr and is right at the limit for human life. It is an extraordinary coincidence that the highest peak on Earth is also about the highest point at which humans can survive unaided. Indeed, so close is Everest to the maximal altitude that we can attain, that minor variations in barometric pressure, such as those caused by the season, can mean the difference between the success or failure of an ascent without oxygen.

Another obvious way to get more oxygen to the tissues would be to increase the oxygen-carrying capacity of the blood. In some animals, oxygen is carried in the blood simply in solution. The amount that can be carried this way, however, is very small and most animals, including humans, use proteins as oxygen carriers. As these proteins are usually coloured they are often called respiratory pigments. In most mammals, haemoglobin is responsible for transporting oxygen. It is composed of four identical subunits, each of which has an iron atom sitting at its heart. One oxygen molecule binds reversibly to each iron atom. Haemoglobin is small enough to be filtered by the kidneys into the urine and is therefore contained within red blood cells, giving them their characteristic colour; red urine is a tell-tale sign of red blood cell disease (unless, of course, you happen to have eaten beetroot recently).

A long-term adaptation to altitude, indeed the first that was recorded, is a marked increase in the number of red blood cells (and therefore also in the haemoglobin concentration). This is triggered by the hormone erythropoietin, which is secreted in response to low blood oxygen levels. Somewhat surprisingly perhaps, this hormone is produced by the kidneys. It appears that expression of the erythropoietin gene, and thus manufacture of the hormone, is switched on by a fall in oxygen. The mechanism is not fully understood but the gene itself (the DNA) is believed to have a control element that directly senses the oxygen concentration in the cell. The increase in circulating red blood cells triggered by erythropoietin begins within three to five days of arrival at altitude and continues as long as the individual remains there. The volume of blood occupied by the red blood cells (known as the haēmatocrit) is around 40 per cent in a lowlander but

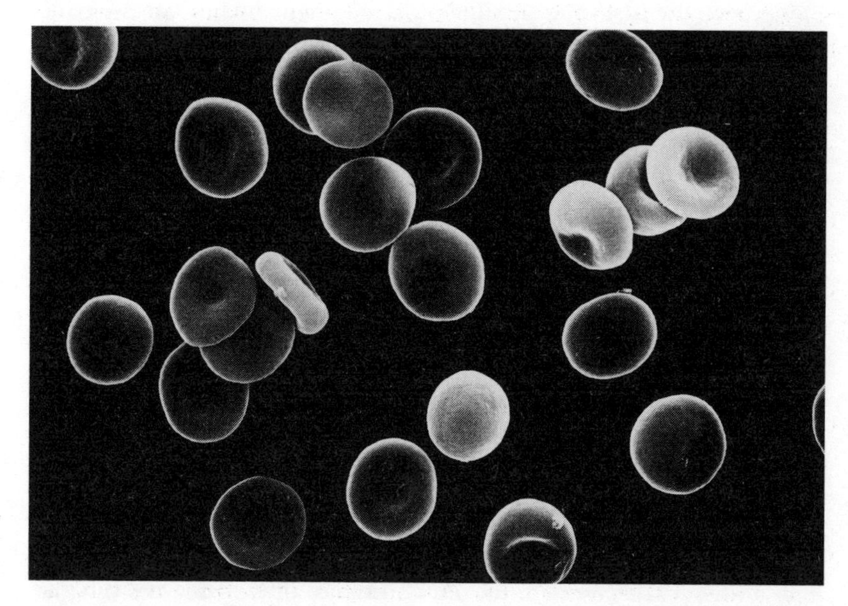

Red blood cells are packed full of haemoglobin. There are around 5000 million of them per millilitre of blood and they contain about 150 milligrams of haemoglobin. They lack a nucleus, have a biconcave disk-like shape, and are so distensible that they can squeeze easily through the finest capillaries. Red blood cells have an average life of 120 days in the circulation and new cells are constantly being made in the marrow of the bones.

can rise as high as 60 per cent following acclimatization. Athletes often train at altitude in order to increase the number of their red blood cells and enhance the oxygen-carrying capacity of their blood – although nowadays some of them breathe air with a reduced oxygen concentration while they sleep, or take genetically-engineered erythropoietin instead (see Chapter Five). People with chronic lung disease, who have difficulty breathing (and therefore suffer from hypoxia), also often have an elevated number of red blood cells, even at sea-level.

Although the increased number of red blood cells enhances the capacity of the blood to deliver oxygen to the tissues, it also produces a concomitant increase in the viscosity of the blood, which makes it harder for the heart to pump the blood around the body. It is currently thought that the increase in the haematocrit is of little benefit (perhaps someone should tell the athletes), a view supported by the fact that llamas and other animals adapted to live at altitude have a similar number of red blood cells to lowland animals. Indeed, if the density of red cells rises too high it can have deleterious consequences. Carlos Monge was the first to notice, in 1925, that some individuals who had lived all their lives at high altitude developed symptoms similar to those of acute mountain sickness. They complained of headaches, dizziness, drowsiness, chronic fatigue and, in some cases, exhibited signs of heart failure or suffered from strokes. They also had haematocrits as high as 80 per cent. Even today, it is possible to observe natives of cities such as La Paz (3500 metres) who have the blue lips and fingernails, and the clubbed fingers characteristic of Monge's disease. These are caused by sludging of the red blood cells in the capillaries, which dramatically reduces the rate of blood flow and thereby also the oxygen supply. Descent to lower levels alleviates the problem and individuals with Monge's disease are faced with permanent exile at sea-level. Why their bodies should lose the ability to adapt to altitude, and why it should be more common in men than women, remains a mystery.

The dramatic increase in the rate and depth of breathing, the renal regulation of the blood acidity, and the reduced sensitivity to the effects of carbon dioxide constitute the body's most important adjustments to high altitude. They account for our ability not only to survive, but also to undertake strenuous exercise, on the summit of Everest without using supplementary oxygen.

Lowland natives who move to the high mountains when adult never attain the level of acclimatization found in people who have lived there all their lives, even if they live there for many years. High altitude natives have much larger, barrel-shaped chests, with proportionately larger lungs; they are also smaller, so that their ratio of lung volume to body size is increased. Their hearts are larger than those of lowlanders, enabling them to pump blood more efficiently around the body, and their lungs and tissues have more capillaries, which facilitates oxygen uptake and delivery. These anatomical adaptations explain why the work capacity of such people is so much greater than that of lowlanders, even when the latter are well acclimatized. Fit young Europeans trekking high in the Himalayas are often astonished (and embarrassed) at the very great loads that are nonchalantly carried by ancient porters or young Sherpani girls – loads that they find difficult to lift, let alone carry for many miles.

The adaptations demonstrated by high altitude natives appear to be partly genetic and partly developmental, because children of lowland races born and raised at high altitude develop larger lungs, but never attain the barrel chest of certain Andean peoples.

Lessons from High Altitude Studies

The story of high altitude physiology is littered with cautionary tales. Physiologists have repeatedly stated that it would be impossible for people to ascend above a given level, only to be confounded as mountaineers proceeded to prove them wrong. Why this was the case provides an illuminating picture of how science actually works.

The first errors arose over the estimation of the barometric pressure on the summit of Everest. Early investigators showed that the barometric pressure varies with the temperature of the air, and rises with increasing temperature (this is because the pressure of a gas depends on the speed at which the gas molecules bombard surrounding objects). With the advent of aviation it became necessary to develop a standard method for calibrating altimeters and, for reasons of convenience, this method assumed a standard temperature at sea-level and a standard rate of decrease with altitude. It therefore did not take account of the effect of seasonal variations in temperature, nor of the fact that the thickness of the atmosphere varies with latitude, being thicker at the equator and thinner at the poles.[9] Consequently, calculations using the standard atmosphere method predicted a lower barometric pressure on

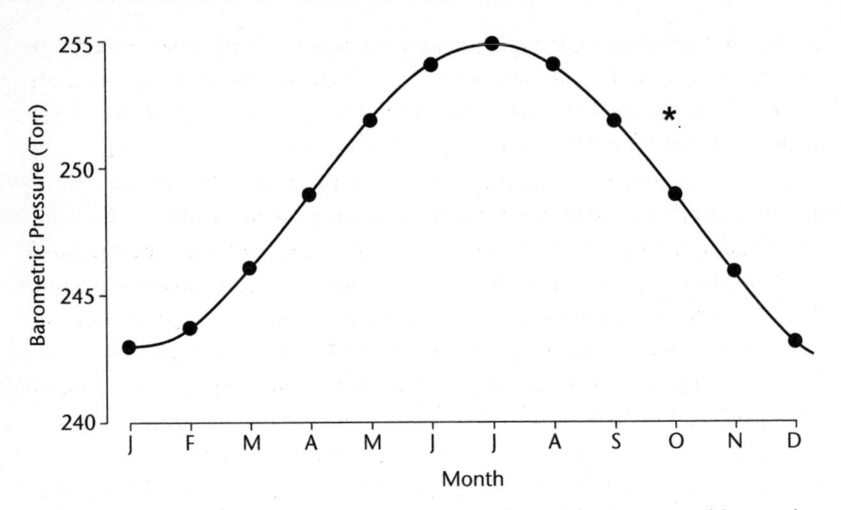

Mean monthly barometric pressure at 8848 metres (29,000 feet), measured by weather balloons released from New Delhi. The asterisk marks a measurement obtained from the summit of Mount Everest (8848 metres) the same year. The barometric pressure varies considerably with season, being higher in the summer when the temperature rises. It is therefore easier to reach the summit of Everest in the summer, when the higher barometric pressure means that the oxygen concentration of the air will be greater. In winter, the lower barometric pressure, and consequently lower oxygen level, is exacerbated by the harsher weather conditions. It was not until 1987 that Sherpa Ang Rita made the first winter ascent without oxygen. He remains the only individual to have done so and it is possible that his success was aided by the unusually fine weather in December that year.

the summit of Mount Everest (236 Torr) than is in fact the case and some scientists concluded that it was unlikely that anyone would be able to survive without supplementary oxygen. The more astute were aware that the estimated barometric pressure was too low but they still did not know what it actually was. It was not until the 1981 American Medical Research Expedition to Everest that the barometric pressure was actually measured, by Dr Chris Pizzo, and found to be 253 Torr. This story illustrates the importance of defining every variable as precisely as possible when making a calculation, and the errors that can arise when such variables are estimated rather than measured. It is also interesting to note that were Everest to be located at one of the poles, the barometric pressure would indeed be too low to survive on the summit without extra oxygen.

Another source of error arose when estimating the oxygen concentration in the lungs on the top of Everest. One of the first comprehensive studies of the effects of long-term adaptation to altitude was carried out by Mabel Purefoy FitzGgerald during an Oxford University expedition to Pike's Peak in Colorado in 1911 led by the distinguished physiologist John Scott Haldane. Fitzgerald studied physiology as an undergraduate at Oxford. In those days women were permitted to take the examinations (a recent concession) but their names were not published in the Class lists nor were they awarded a degree. Mabel distinguished herself by obtaining First Class marks. She remained at Oxford working within the Physiology Department, during which time she carried out a number of studies on respiration. In 1911, together with Haldane, Gordon Douglas (another eminent physiologist) and others, Mabel journeyed to Pike's Peak, at 4302 metres one of the highest mountains in the United States. Their aim was to study the effects of altitude on the human body (their own, in fact). This was not an unduly arduous expedition: a steam-powered funicular railway carried them straight to the top of the mountain, which was crowned with a small hut known as the Summit House. Here the men stayed in relative luxury. Mabel,

Mabel FitzGerald with other members of the Anglo-American Pike's Peak expedition of 1911 (from left to right: J.S. Haldane, M.P. FitzGerald, E.C. Schneider, Y. Henderson and C.G. Douglas)

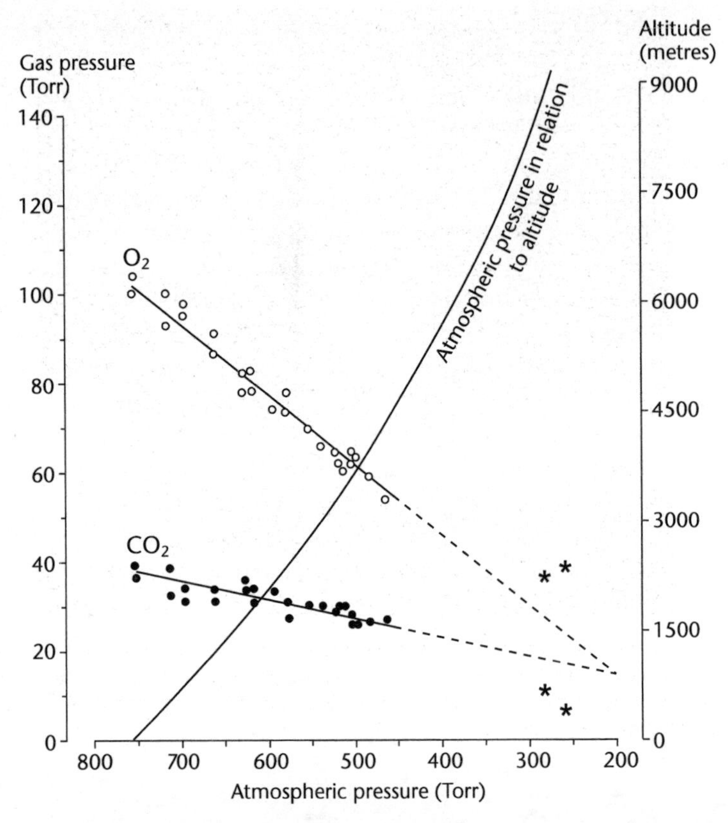

The relationship between atmospheric pressure and carbon dioxide (CO_2), or oxygen (O_2), concentration in the lungs of acclimatized people is linear up to about 5500 metres (18,000 feet), where the atmospheric pressure is 400 Torr. After that, the relationship deviates from linearity because the increase in the rate and depth of breathing blows off more carbon dioxide from the lungs and thus provides more room for oxygen. The dashed lines indicate the levels of carbon dioxide and oxygen predicted if the relationship is assumed to remain linear. The circles indicate the data obtained by Mabel FitzGerald on the Pike's Peak expedition and elsewhere. The asterisks are data obtained by Dr Chris Pizzo on the summit of Everest (see illustration opposite).

however, was excluded – possibly because of the difficulty of organizing the sleeping arrangements. She was despatched on a mule to lower altitudes to examine the haemoglobin content of the blood, and the carbon dioxide concentration in the expired air, of the local mining population.

Her efforts were well rewarded. She confirmed earlier observations that the haemoglobin content of human blood, and thus the number

Dr Chris Pizzo taking an alveolar gas sample on the summit of Everest during the American Medical Research expedition in 1981. After the struggle to get to the top, and a pause to admire the view, he set to work collecting samples of alveolar air from himself. Physiology can sometimes be a physical as well as an intellectual challenge!

of red blood cells, is increased in acclimatized individuals. Her data also showed a strikingly linear relationship between altitude and the partial pressure of carbon dioxide in air expired from the alveoli. When this relationship was extrapolated to 8848 metres, the height of the summit of Mount Everest, the partial pressure of alveolar carbon dioxide was predicted to be around 15 Torr.[10] At this level of carbon dioxide, the partial pressure of oxygen in the lungs works out as approximately 20 Torr, well below the limit for human survival. For many years, this led to the erroneous idea that it would not be possible to reach the summit of Everest without supplementary oxygen. In retrospect, it is easy to see why this mistake occurred. It turns out that above 5500 metres the relationship between altitude and the partial pressure of carbon dioxide in the alveoli is no longer linear, because of a dramatic increase in respiration. Consequently, the partial pressure of oxygen in the alveoli on the top of Everest is much higher than predicted (35 Torr rather than 20 Torr) and it is indeed possible to survive, as many climbers have demonstrated. The lesson is that it is always risky to extrapolate beyond one's data (Mabel's stopped at 4270 metres), because there is no guarantee that the relationship will remain the same.

Mabel disappeared from scientific life around 1920. Many years later, she was discovered to be living in Oxford, just across the University Parks from the Physiology Department, and in 1972, when she was a hundred years old, the University of Oxford at last awarded her the degree that she had earned years earlier.

High Life

Although the low oxygen concentration is the primary difficulty for someone standing on the top of a high mountain, other factors, such as cold, dehydration and sunburn, also pose problems. Solar radiation is unusually intense because the thinner air provides less screening, and as this is exacerbated by reflections from snow and ice it can lead to severe sunburn. The humidity also decreases at altitude, as the reduction in temperature and atmospheric pressure means that the water vapour content of the air is reduced. Dehydration, which is aggravated by the increased respiration, is therefore a problem and it is essential to drink a lot in order to replace the water that evaporates from the lungs during breathing – not always an easy task when you must either carry the water with you or take enough fuel to melt snow. Most serious of all is the cold. The temperature falls by approximately 1°C for every 100-metre increase in altitude because the insulating effect of the atmosphere is less, due to the thin air, and consequently more heat escapes into space by radiation. The decrease in temperature is compounded by high winds, which produce a further 'wind-chill' factor. Several mountaineers have lost the tips of their fingers or their toes to frostbite; on the 1988 expedition up the notorious Kangshung face of Everest, for example, Steve Venables lost three and a half toes while Ed Webster had to have three toes and the last joints of eight fingers amputated. Others have died. Why this happened, and how the body copes with extremes of cold, is the subject of Chapter Four.

Taking the Plunge

When I arrived in Puerto Rico I had never even opened my eyes underwater, let alone dived to the bottom of the sea. All this was to change. By the time I left, I had made my first scuba dive on a coral reef and been hooked for life.

My destination was a research institute in San Juan, the capital of Puerto Rico. Housed in an old stone fort built into the city wall, on a cliff high above the sea, its scientists were engaged in studies of how nerve cells work, whether there was any connection between the nervous and immune systems, and of the rare and beautiful creatures that lived on the island and in the surrounding seas. In addition to the laboratories, there were several dormitories for visiting scientists like myself. Most of my time was spent at the institute, but on two occasions I was taken to visit the coral reefs that border the island.

On my first trip, my friends kitted me out with gas tank and demand valve and walked by my side in the shallow water that fringed a coral atoll while I became accustomed to the equipment. Absorbed in contemplation of the small fish scurrying across the sand, I abruptly found myself struggling for breath, a difficulty not aided by my companion who kept trying to push my head below the surface. I was indignant – the gas had run out. 'Never mind,' was his response. 'We'll go snorkelling.'

And so I was introduced to Paradise.

My hair floated horizontally around my head, lazily drifting this way and that as I moved, in a slow-motion underwater ballet. A myriad fish clothed in brilliant jewel-like colours swarmed around me. Small vivid yellow-and-blue striped ones, with flattened bodies that help them to vanish when viewed end-on. Schools of others, twisting and turning in synchrony as they wove their way through the crevices of the reef. Fish with black and purple blotches; with eyes staring from their tailfins; with dorsal fins trailing behind them like streamers; fish decked out in silver and blue or wearing gaudily-coloured patchwork coats. A shoal of large groupers, severe in grey and brown, with lugubrious faces sailed ponderously past. A coral-coloured fish with rose and olive spots dived for cover. In my hand I held a small plastic bag containing crumbled pieces of cheese; if I opened it a crack, a cloud of eager fish suddenly surrounded me, attracted by the scent. How strange that fish should have a passion for cheese. Something kissed my foot and I looked down to see a small fish with rubbery pouting lips nibbling my ankle. I was so engrossed in this exuberantly beautiful underwater world that I hardly noticed that I had to surface for air from time to time.

Three days later, the day dawned grey and overcast, hardly an auspicious start for my first scuba dive. My companions kept reiterating instructions as we drove along. 'Keep close to us . . . any problems, head for the surface . . . remember, breathe out when you ascend . . . don't get cold.' I listened diligently. By the time we reached the dock, it was spitting with rain. We bounced over the waves towards the reef, anchoring in the lee of a small tree-covered island. The boat rocked up and down with the swell, as thunderclouds gathered overhead. I peered over the side, straining to see the reef, but the visibility was poor because the previous day's storm had kicked up a lot of sand. I lowered myself carefully into the murky water, manoeuvred the heavy gas bottles on to my back and clipped on the weight belt. I was expecting to sink, but I seemed surprisingly buoyant.

'Don't worry,' I was told. 'Just grab hold of the anchor chain and walk steadily down it. We'll be with you directly.'

I tried to follow their instructions, but no matter how hard I endeavoured to pull myself downwards, hand over hand, using the anchor chain, I mysteriously kept bobbing up to the surface. And I didn't seem to be getting any air out of my tank. One of my companions noticed my difficulties.

'What's the matter? Are you afraid?'

'Yes' I said softly, for, suddenly, I realized I was terrified. All those cautionary stories about the necessity of breathing out during an emergency ascent in order to avoid bursting the lungs had had a profound effect.

'All right,' he replied, 'get into the boat. You can't dive if you're scared.'

'But . . .'

'No, sorry. Get back into the boat.'

I hauled myself miserably over the side, slithering forwards on my stomach like a beached seal. My friends gathered themselves together, nodded to each other, and tilted backwards over the side of the boat. With a casual wave they were gone. I sat in the cockpit in tears, the rain hissing down into the sea around me. I felt excluded – a feeling not helped by the realization that it was all my own fault, for I had been given the opportunity but had simply been afraid to grasp it.

I was awakened from my reverie by a shout. A black form emerged dripping from the sea, removed its mouthpiece and spoke. 'Are you ready now? I've got an hour of air left. D'you want to come and see the reef?'

This time it was easy. I had no problem sinking, nor did I experience the same difficulty breathing. Now I know that previously, in my fear, I had filled my lungs with air, but had forgotten to breathe out. So I had been far too buoyant; and I couldn't breathe, not because there was no air in my tank, but simply because my lungs were already full.

I sank below the surface and the reef opened up before my eyes. For someone trained as a zoologist, as I had been, this was a breathtaking experience. I could spend hours just watching one small patch of reef – and with my own air supply it was possible to do so. Polychaete worms, endlessly extending and retracting their bodies, opening and closing the feathery flower-like fans at their tips to sieve the water for the minuscule life-forms on which they feed. A small crab, barely detectable as it sat motionless amidst them, just the glinting of its eyes giving it away. Sea anemones, their tentacles waving slowly and sedately in the current until stimulated by some chance encounter to curl around the unfortunate victim. A parrot fish, resplendent in orange and white, protected by their arms. And the coral itself, myriads of individual polyps looking like flowers but really animals, strung together by protoplasmic highways that course through the protective carapace the colony secretes. Within their cells, the coral polyps harbour unicellular blue-green photosynthetic algae that fix atmospheric carbon dioxide, so supplying nutrients to its host but dictating a life in the light upper reaches of the sea. A marriage of plant and animal, partners for life, that is of importance for the carbon cycle of the earth, for the coral polyp locks away carbon dioxide by converting it to calcium carbonate and laying it down to form the reef. Colonies of tunicates decked out in yellow and deep mauve; vigorous and with a well-developed nervous system when young, in middle age they give up an active existence, anchor themselves to a rock and never move again. In this sessile state, they lose their nervous system for it is no longer needed. A dire warning, perhaps, to those of us who take insufficient exercise!

2

LIFE UNDER PRESSURE

'They that go down to the sea in ships,
that do business in great waters;
These see the works of the Lord
and his wonders in the deep.'

Psalm 107

The legendary descent of Alexander the Great into the straits of the Bosphorus in a glass barrel

WHEN VIEWED FROM space, the Earth is a beautiful iridescent blue ball, hanging in the darkness. Seeing our planet in this way, we realize that we live on a world of water. The landmasses that man occupies cover but a small part, roughly a quarter, of the globe's surface, and most are concentrated on one side of the planet. Even if we have never set foot on their shores, the oceans touch our lives, for here is where weather is woven and hurricanes are born. Changes in the ocean currents, most famously the El Niño, extend their influence across the globe, creating drought and famine in some areas and torrential rains in others. The England I live in is a green and pleasant land with mild weather and long seasons because of the Gulf Stream which warms its shores. Yet, despite their vast size – 260 million square kilometres of the planet's surface – and their importance, we still know little about the oceans. Most of our knowledge is confined to the shallow shelves at the edges of the continents and even today, when men have walked on the surface of the moon, the deepest regions of the sea remain largely unexplored.

At 10,914 metres (35,800 feet) the Mariana Trench in the Pacific Ocean is the deepest part of the ocean floor, so deep, in fact, that you could drop in Mount Everest and still have 2000 metres to spare on top. It has rarely been visited by man. Even the average depth of the sea, around 4000 metres, is too far for us to reach except in a submersible. Yet, perhaps because of this very inaccessibility, people have always been fascinated by the abyss. Stories of mythical creatures that live far

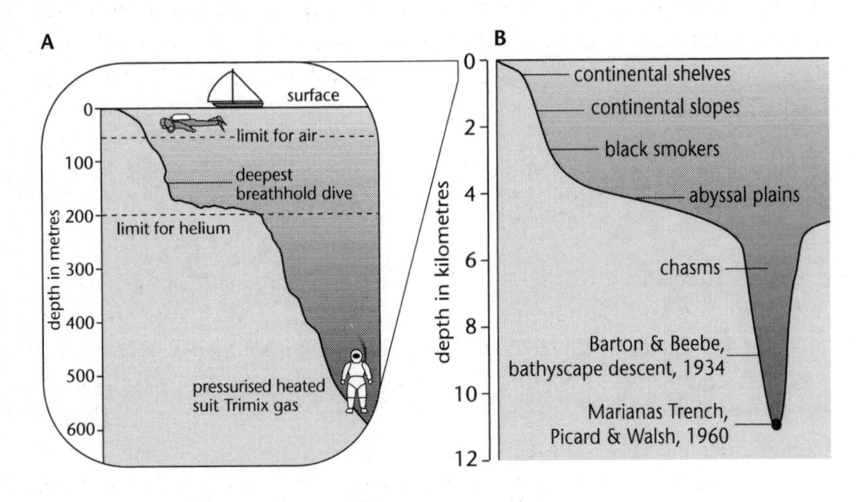

The Continental shelves which surround the landmasses form fertile plains lit by the sun, that are rich in animal and plant life. They slope gradually downwards to a depth of 200–300 metres. The sea floor then angles more steeply until it reaches the abyssal plains, 3–6 kilometres (or 2 miles) deep, where the bottom is covered with soft mud (ooze). These are interrupted in places by deep chasms, such as the Mariana Trench in the North Pacific (10,914 metres deep) and the Puerto Rico Trench in the North Atlantic (8384 metres deep).

below the waves abound in many cultures. Here lies Poseidon's palace and the mermaids' home; the Kraken[1] slumbers; and Leviathan (a ferocious monster in Phoenician mythology) took refuge when vanquished by the Creator. The truth, as so often happens, is yet stranger still. The scientific world was astonished in 1938 when a living coelacanth, previously known only from fossil records, was discovered; and despite never having been observed alive, giant squid, with arms up to 18 metres long, do exist, for their bodies have been dredged from the depths and their beaks are found in the stomachs of whales. Even more extraordinary are the bacteria described in Chapter Seven, which live around the 'black smokers' of the mid-ocean ridges at temperatures of over 100 °C and pressures of over 1000 atmospheres.

Although people do not live permanently underwater, some of them, such as divers attached to North Sea oil rigs, may spend a significant part of their lives there. Many thousands more go scuba-diving, snorkelling or skin-diving on holiday. What are the problems they encounter and how deep is it physically possible for them to go? This chapter looks at the way in which a knowledge of human physiology has been intimately linked with our increased ability to spend time at depth and it asks why, once again, animals seem to do it with so much less effort.

The Physics of Pressure

Apart from lack of air, the main difficulty experienced by a diver is the increased pressure. The deeper one descends in the ocean, the more the pressure increases, because of the weight of the water pressing down on top. As water is some 1300 times heavier than air, over the same vertical distance, the pressure difference is much greater in water than in air; the pressure decreases by two-thirds between sea-level and the top of Everest (8848 metres), but increases 885 times when descending from sea-level by the same amount. The pressure at the bottom of a column of liquid is determined by the height of the column, the density of the liquid and the force of gravity. In sea water, the pressure increases by approximately one atmosphere for every 10 metres (33 feet) that you descend. Divers usually measure pressure in units of atmospheric pressure, known as bars; at a depth of 30 metres, the pressure is 4 bar, being the sum of the pressure at the surface (1 bar) and that underwater (3 bar).

The volume of a gas varies with pressure. Robert Boyle (1627–91) described this phenomenon in a famous law, that he formulated in his Oxford laboratory, which lies not far from my own. He showed that (at a given temperature) the product of pressure and volume is always constant: in other words; pressure × volume = constant. Thus, at a depth of 30 metres, where the pressure is four times atmospheric, the volume of a gas will be reduced to a quarter of that at the surface. As we shall see later, this compression of a gas at depth, and its expansion when the pressure is released on ascending to the surface, has profound implications for the diver.

The Earliest Divers

Diving for food, salvage and military purposes is an ancient tradition. One of the earliest references to diving is in the *Iliad*, where the Greek warrior Patroclus sarcastically compares the way in which Hector's charioteer tumbles from his chariot when hit by a sharp stone with a man diving for shellfish. Other early Greek writings make reference to sponge-divers and their use of lead weights and ropes to speed their ascent and descent. Ornaments decorated with mother-of-pearl suggest that people may have harvested shellfish as long ago as 4500 BC in Mesopotamia, and women have been diving professionally for pearls, seaweed and shellfish in Japan and Korea for at least 2000 years, for they are mentioned in the *Gishi-Wajin-Den*, which is believed to have been written around 250 BC. We also know that divers were trained and employed in offensive naval warfare by the Greeks as early as 400–333 BC. The most famous was Scyllias who, according to Herodotus, was engaged by the Persians to recover treasure from sunken ships, but then deserted to the Greeks and helped them win a battle against the Persians by providing them with valuable information about the enemy fleet and by cutting enemy cables underwater.

Water spiders use diving bells, made of a net of silken threads anchored by guy ropes to the submerged stems of aquatic plants. They fetch air from the surface to stock the bell by trapping a small bubble between their hind pair of legs: it may take several journeys to fully furnish their air supply. The water spider is a hunter and lurks in her silken air-filled dome with her front legs protruding into the water, waiting to pounce on any unsuspecting prey that passes.

The use of diving bells and sealed diving vessels is also very old. In crude form, the diving bell was invented in the sixteenth century, but it was not really until the hand pump was invented, by the German Otto von Guericke in 1654, that a method of replenishing the air supply was available and the use of diving bells became a practical possibility. The principle of the diving bell is easily demonstrated by taking an empty jam jar and inverting it in a basin of water. As you will observe, the presence of the air excludes the water. One problem with this simple form of the diving bell is that if it is not kept upright, air will escape from the edge of the bell and water will seep in to replace it (try tilting the jar). Another difficulty is that the volume of air contained in the bell will decrease as the pressure increases, in accordance with Boyle's law: at a depth of 10 metres, for example, the air will be reduced to half its original volume. Thus it is necessary to supply the diving bell with air via a hand pump from the surface.

Diving suits were first developed for marine salvage. Among the early pioneers were two brothers, John and Charles Deane, who set up business as 'submarine engineers' in the Portsmouth area of England around 1832. This came about in a highly unusual way. While trying to rescue a group of horses from a burning barn, they had the bright idea of using a helmet from a suit of amour, supplied with air via a hose and a hand pump, as a breathing apparatus. It was so successful that they patented the device for fire-fighting. They soon realized that it would also be suitable for diving and by 1828 they had perfected a diving apparatus consisting of a heavy open helmet, balanced on the diver's shoulders, which was supplied with compressed air via a leather hose from a pump on the deck of the support ship. Providing the diver's head was upright, the helmet acted as a portable diving bell and water was prevented from entering at the bottom of the helmet by virtue of the air that was supplied from above.

This outfit was successfully used for a number of years to dive to depths of up to 10 metres for periods of up to thirty minutes. It had, however, one obvious disadvantage: if the diver fell over, his helmet filled with water and he was likely to drown. The introduction of a sealed diving suit, in which the helmet was tightly clamped to a waterproof suit, solved this problem but it also introduced a new one. The air supplied from the surface now filled the diver's suit as well as his helmet. If the diver descended too rapidly, or unexpectedly, his assistant might not be able to increase the air pressure to that of the surrounding

Contemporary painting of Mr Deane, self-styled submarine engineer, at work on the wreck of HMS *Royal George* (of 108 guns) which sank off the Isle of Wight in England in 1782. He is 'equipped in his newly invented Diving Apparatus' and is 'engaged in taking off one of the hoops of the bowsprit in August 1832'.

water sufficiently quickly and consequently the volume of air within the suit would decrease (recall that pressure × volume = constant). The diver's head was protected by his copper helmet, but the external water pressure would squeeze the suit painfully around him, sometimes compressing his chest so much that it damaged his lungs. Often, the diver would feel as if his whole body was being forced into his helmet. In the worst case, when the non-return valve between the air-hose and the diver's dress failed from the pressure, 'his blood and much of his flesh go up the hose pipe and all that is left in the dress is some bones and some rags of flesh'.

The amount of air in the suit determined the buoyancy of the diver and could be reduced slightly to aid his descent, or increased to help him rise. This was regulated by the diver, who adjusted the rate at which the air, which was supplied at a constant rate, escaped through a one-way valve on the side of the helmet. While too little air in the suit

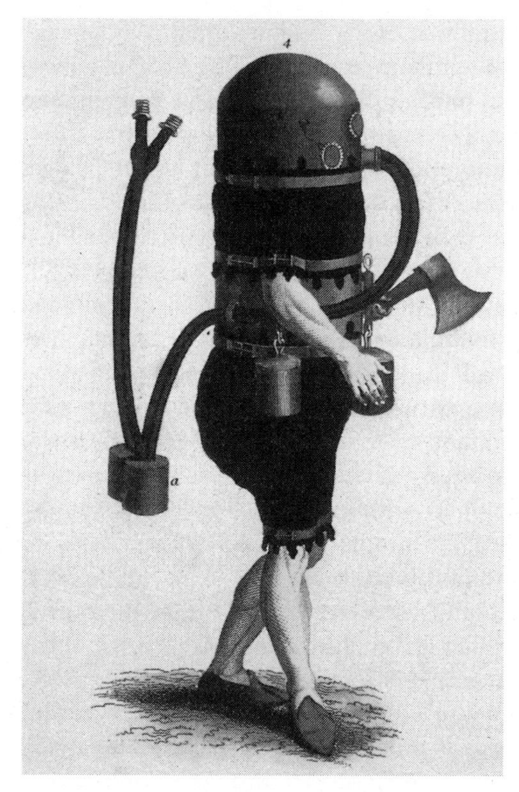

One of the more extraordinary types of early diving suit

caused divers' 'squeeze', too much air in the suit was also a problem. If the legs of the suit filled with air, as sometimes happened when he was crawling around on the bottom, the diver might suddenly find himself turned upside down. In this position, the excess air could not easily escape and he was blown helplessly to the surface. Nevertheless, these difficulties could be avoided by an experienced diver and support team. And, increasingly, divers were required, not only for military and salvage operations but also in construction work.

In the middle of the nineteenth century, the invention of the steam locomotive heralded the dawn of the great railway age. The landscape began to change as railways were laid from one end of the country to the other, established towns swelled to previously unimaginable proportions and new ones were built. Suddenly, people and goods could be shuttled about at great speed and in large amounts. To the people of that time the sudden increase in the ease of communication must have appeared like the growth of the Internet today. What began in Britain spread throughout Northern Europe and by 1850 a railway network connecting the main towns of France, Germany, Belgium and Britain had been established. The engineers were men of spirit – they tunnelled through mountains and under rivers and threw bridges across large rivers and estuaries. It was during the course of constructing these bridges and tunnels that they found their workers succumbed to a disease that was soon known as diver's palsy, or caisson sickness.

Caissons, introduced in about 1840 by the French engineer Triger, were used for constructing foundations for the piers of bridges over rivers. The caisson is a watertight double-skinned steel tube, open at the bottom, that ultimately forms part of the pier itself. The inner tube provided access for the workmen and removal of material, and was filled with compressed air to keep out the water, while the concentric outer walls were gradually filled with concrete from above as the caisson sank into the river-bed. For simpler types of work on river beds or harbour bottoms, traditional diving bells were used. These were lowered to the bottom with the labourers sitting inside them and, being supplied with compressed air, enabled the men to work under dry conditions. When it was necessary for men to move around freely underwater, individual divers were employed. Compressed air was also pumped into tunnels to prevent water seeping through porous rock into the tunnel during its construction. Thus, many construction workers on bridges and tunnels worked in an atmosphere of compressed air, often for as long as eight hours each day.

Right from the start it was noted that soon after returning to atmospheric pressure, tunnel and caisson workers frequently became ill. Commonly, they complained of itchy skin. Less frequently, they experienced severe pains in the limbs that prevented them from extending their joints – which is why the workmen called such attacks the 'bends'. The bends never happened while actually under pressure but only following return to atmospheric levels; as Pol and Watelle remarked in the first medical account of caisson disease *'on ne paie qu'en sortant'*.[2] The risk of an attack and the severity of the symptoms increased with the pressure and the duration of exposure, and divers, who were invariably subject to greater pressures, suffered more than caisson workers. In the worst cases, on return to the surface the victim felt faint, rapidly became paralysed, lost consciousness and died, all within a few minutes.

Bubbles in the Blood

The cause of the bends was established by the French scientist Paul Bert in 1878. He showed that they result when a diver or caisson worker breathing compressed air is decompressed too quickly and gases that have dissolved in the blood and tissues are liberated in the form of bubbles, producing blockage of the blood vessels. When a gas is breathed under pressure, more of it dissolves in the body fluids: an additional litre of nitrogen, for example, for every 10 metres of descent (the process takes time, see later). The presence of the additional gas in

Why Don't Sperm Whales Get the Bends?

Many marine mammals dive to far greater depths than humans can tolerate. A dead sperm whale was once found at a depth of 1134 metres with its lower jaw entangled in a transatlantic cable. Elephant seals are even more prodigious divers, for the deepest dive that has been recorded is a staggering 1570 metres, where the pressure is more than 150 times that at the surface. This is way beyond the limit for humans. Moreover, seals may dive repeatedly without ill effect. Indeed, the elephant seal should more correctly be called a surfacing animal than a diving animal, for it spends more than 90 per cent of its time at sea underwater; one animal was observed to spend no more than six minutes on the surface during forty days at sea. So why don't seals and sperm whales get the bends?

The answer is that marine mammals have developed ways of reducing the amount of nitrogen that dissolves in their tissues. Unlike humans, seals and whales breathe out before diving. This limits the amount of air they carry with them and at a depth of about 50 metres the air-sacs in the lungs are completely collapsed, so that no further gases enter the bloodstream. The pressure at depth causes the whale's lungs to collapse completely, forcing all the air into the upper airways which are supported by circular discs of cartilage and are therefore less compressible. Blood flow to the lungs is also markedly reduced. These adaptations ensure that no gas enters the bloodstream from the whale's lungs during a dive. This means that little extra nitrogen dissolves in the body fluids, so there is no danger of bubble formation on decompression when the animal ascends.

the body fluids and tissues is not a problem so long as it remains in solution. It is the rate at which the dissolved gas can be eliminated on decompression that is the difficulty. When a diver ascends slowly, the extra gas dissolved in his blood is simply excreted by the lungs and causes no difficulties, but if his ascent is too rapid the rate at which dissolved gas can be eliminated via the lungs is exceeded, so that the tissues and blood become supersaturated with gas. At some point, this gas suddenly comes out of solution as bubbles.[3] Everyone who has opened a bottle of sparkling water (or champagne) will be familiar with this phenomenon – once the pressure is released, the dissolved gas expands in a fizz of bubbles. This is a much more dramatic affair if the cap is removed suddenly (rapid decompression) than if the gas is allowed to leak out very gradually by slowly untwisting the cap. The dissolved gas in mineral water (and champagne) is carbon dioxide, but for a diver breathing compressed air nitrogen is principally responsible for bubble formation, because the carbon dioxide concentration is very low and oxygen is rapidly utilized by the tissues.

Bubbles in the blood create serious problems. Once formed, they tend to grow in size, as more gas diffuses into them. Consequently, they may become large enough to block the finer blood vessels and prevent blood flow to the tissues, leading to a lack of oxygen and nutrients that can cause cell death. The presence of the gas bubbles may also activate blood cells that are primed to respond to air – such as the platelets, which are involved in blood-clotting. In addition, tissues may be damaged if gas bubbles form within them, since this may deform, or force apart, the cells and disrupt their function.

Divers have coined special names to describe the various symptoms associated with the formation of bubbles in different tissues. The 'chokes' refers to the breathing difficulties experienced when large bubbles become trapped in the lung capillaries, reducing the surface area available for gas exchange and producing a feeling of breathlessness. The 'staggers' results from bubbles in the vestibular apparatus of the inner ear, which is normally involved in the control of balance. Bubbles in the knee and shoulder joints, the most common sites for decompression sickness, produce the 'bends'. When they form in the spinal cord, bubbles cause pins and needles or paralysis and in severe cases can lead to degeneration of the nerve fibres. Cerebral bends – bubbles in the brain – are associated with visual and speech disturbances and may be fatal.

There is a story, possibly apocryphal, that when one of the first tunnels was being built under the river Thames, the directors decided to celebrate reaching the halfway mark by hosting a lunch in the tunnel. Because it was not yet completed, the tunnel was still filled with compressed air and the guests had to take their lunch 'under pressure'. Somewhat to their chagrin, the champagne did not 'pop' when it was opened, nor were there any of the usual bubbles, for the pressure in the champagne bottle was the same as that in the tunnel. Nevertheless, despite the flat taste, they all drank the champagne. The bubbly lived up to its name later, however, when the directors and their guests returned to the surface, and to atmospheric pressure, and the champagne they had drunk began to fizz!

The Importance of a Slow Ascent

Compressed-air workers soon discovered for themselves that their symptoms could be relieved by returning to the higher atmospheric pressure in which they had been working. This led Sir Ernest Moir to suggest a recompression chamber should be used for the treatment of caisson disease. It was first employed around 1890, in connection with the construction of the Blackwall Tunnel under the river Thames and the East River Tunnel in New York, and it proved strikingly successful. But it often took many hours to decompress someone suffering from the bends. Clearly, what was needed was some method of preventing an attack in the first place. As a result of Paul Bert's work, the solution was obvious: a diver or caisson worker must ascend (or decompress) sufficiently slowly for the dissolved gas to be expelled by the lungs. The difficulty lay in determining what rate of decompression was safe. By 1906, the problem had become so acute that Professor John Scott Haldane of Oxford University, a physiologist already well known for his work on respiration (see Chapter One) was asked by the British navy to sort it out.

Together with Lieutenant G.C.C. Damant and Professor A.E. Boycott, Haldane carried out a series of experiments at the Lister Institute in London using a large steel chamber in which the pressure could be easily controlled. Working with goats, they found that there were no ill effects of suddenly decompressing an animal from 6 atmospheres to about 2.6 atmospheres. If, however, they reduced the pressure by the same absolute amount, but in this case from 4.4 atmospheres to 1 atmosphere (i.e. sea-level), the result was very different. Only 20 per cent of the animals were unaffected; the remainder all suffered the bends, some of them so severely that they died. By a series of trial-and-error experiments they discovered that it was safe to halve the absolute pressure rapidly, but that subsequently the pressure had to be reduced far more slowly. Thus, the depth limit for a dive which required no decompression was 10 metres (2 atmospheres). In the time-honoured tradition of experimental physiologists, the team next repeated the experiments on themselves, fortunately without ill effect. The final stages of the experiments were conducted in the sea off the Isle of Bute on the west coast of Scotland, from the naval vessel HMS *Spanker*. Haldane made it a family holiday and allowed his thirteen-year-old son Jack, who was later to develop a great interest in respiration himself, to make a dive of about 12 metres.[4]

Haldane realized that the rate at which nitrogen dissolved in the body's tissues was variable. Fat cells, for example, have a high storage capacity whereas brain cells store little nitrogen (parenthetically, this means that women and fat people require longer for decompression than the average man). Moreover, the rate at which nitrogen accumulates depends on the blood supply to the tissue, and occurs more slowly in tissues with low perfusion rates. As a consequence, it takes over five hours to saturate the human body with nitrogen. On decompression, the dissolved nitrogen must be removed by the blood supply. The rate at which it can be safely eliminated depends on the storage capacity and perfusion rate of the different tissues and, roughly speaking, it takes as long to clear the gas as it does to accumulate it. This means that the best course for a diver is to descend rapidly, spend a limited time on the bottom and then ascend slowly to the surface in stages.

The rapid descent recommended by Haldane and his colleagues contradicted previous practice, but made sound physiological sense as the shorter the period spent at depth, the less gas dissolves in the body. They also specified that the first part of the ascent should be rapid, taking the diver to about half the absolute depth of his dive, a level which they knew from their experiments was quite safe. Thereafter, the diver should ascend slowly, stopping for a fixed time at different levels to allow a gradual period of decompression. The reason for this staged decompression is that the increase in the volume of a gas is the same, regardless of whether the pressure is reduced from 8 to 4 atmospheres or from 2 to 1 (remember that pressure × volume = constant, so that halving the pressure will double the volume). The great advantage of their protocol is that the diver can ascend rapidly, without harm, until the pressure is halved, which allows him to spend longer decompressing at shallower depths. As Haldane himself remarked: 'uniform decompression . . . is needlessly slow at the beginning and usually dangerously quick near the end.'

By 1908, Haldane and his team were able to supply the Royal Navy with a detailed set of decompression tables that specified exactly how long the diver should stop at each depth during decompression, following dives of different depth and duration. After the introduction of these tables, the incidence of the 'bends' dropped precipitously and they were only seen when the diver – for whatever reason – decided not to stick to the guidelines and ascended more rapidly. Not everyone was immediately convinced of the benefits of Haldane's work. As he commented some ten years later: 'It is unfortunate that stage decompression cannot be introduced in some countries on account of antiquated state regulations enjoining decompression at a constant rate, or even decompression starting very slowly and increasing in rate as atmospheric pressure is approached.' Fortunately, the results of his work spoke for themselves and Haldane's method is now routinely used. Tragedies still occur, however, when accidents mean that decompression guidelines are not adhered to. One well-publicized disaster was that of Chris and Chrisy Rouse, a father and son team with considerable diving experience, who died of decompression sickness in 1992 while exploring the wreck of a German U-boat.

It is instructive to compare the decompression time previously employed by caisson and tunnel workers with that suggested by Haldane and his team. Caisson workers were usually exposed to a pressure about three times that of atmospheric (i.e. 3 bars), and decompressed within ten minutes or less. By contrast, after three hours spent working at this pressure, a total decompression time of ninety minutes was recommended by Haldane. It is hardly surprising that so many caisson workers suffered from the bends.

Divers should also avoid flying for some time after diving, because the pressure in an aircraft is less than that at sea-level (see Chapter One) and the further depressurization may cause bubble formation. It is recommended that divers should not fly for twelve hours after a single dive and for even longer after multiple dives or dives that involve decompression stops on the ascent. Holiday-makers unfamiliar with the problems of decompression may suffer from the bends if they spend their morning scuba-diving and then catch a plane home in the afternoon. Military pilots flying unpressurized aircraft may even get the bends if they ascend too rapidly to high altitudes from sea-level.

Skin-diving and the Bends

Skin-divers who descend to great depths in a single dive do not suffer from the bends as they remain submerged for so little time that insufficient nitrogen dissolves in their body fluids to cause a problem on ascent. Repeated deep dives are a different matter, however, as Dr P. Paulev of the Royal Danish Navy discovered to his cost. In the early 1960s, he made around sixty two-minute-long dives, at intervals of between one and two minutes, in a submarine escape-training tank with a depth of 20 metres. About thirty minutes after the last dive he developed an intense pain in his left hip. He chose to ignore it, but some two hours later he experienced severe chest pains, blurred vision, paralysis of his right hand and breathing difficulties. He was discovered in a state of shock by a colleague, who quickly placed him in a compression chamber and recompressed him to a pressure of 6 atmospheres. The symptoms rapidly disappeared. Subsequent decompression took over nineteen hours, but Paulev was fortunate for he completely recovered and later wrote an account of his experience.

The pearl-divers of the Tuamotu archipelago in the South Pacific suffer from a condition called *taravana* which is very reminiscent of that described by Paulev. The word *taravana* translates as 'to fall crazily', and the symptoms range from visual disturbances to loss of consciousness. Occasionally, the victim suffers paralysis or even dies (unlike Paulev, the Tuamotu do not have a recompression chamber to hand). One visitor remarked, 'Go ashore at a place like Barhein island and the greatest population in sight seems to be the burying place for dead skin-divers.' *Taravana* is a much feared and frequent disease. On one day alone, 47 out of the 235 divers working sustained symptoms, some of which were quite severe, for six individuals were paralysed and two died. Fortunately, not all days are so dramatic, but the attrition rate is still rather high.

Although the aetiology of *taravana* remained mysterious for many years, the work of Paulev, and others subsequently, suggests it is most likely to be a form of decompression sickness. The Tuamotu pearl-divers push their bodies to the limits for they dive to depths as great as 40 metres (5 bar), and each dive lasts around two minutes. They make between six and fourteen dives an hour, remaining at the surface for only four to eight minutes between dives. This is probably too short a time to allow all the nitrogen that dissolves in their tissues during the dive to be off-loaded, so that it accumulates with every dive and eventually leads to decompression sickness on ascent (*taravana* never occurs at depth but only on surfacing). Individuals who make many dives at short intervals are most likely to succumb. Interestingly, in the nearby island of Mangareva, where *taravana* is unknown, tradition dictates that the diver spends at least ten minutes on the surface between dives.

On Entering the Water

Decompression sickness is not the only difficulty that faces the diver. Even just immersing the body in water up to the neck causes physiological changes. When you are standing upright on the seashore, there is a pressure gradient down your body due to the force of gravity, which causes the blood to pool in your legs. If you now immerse yourself in the sea up to your neck, this effect is counteracted by the external pressure of the water so that about half a litre of blood shifts upwards from the legs to the chest, distending the great veins and the right atrium of the heart and increasing your cardiac output. One consequence of stretching the atrial wall is that it alters the level of two hormones that influence water uptake by the kidney, and thereby stimulates urine production. This explains why you so often, and so annoyingly, need to pee just after entering the water.

Even immersing your face in water produces a physiological response: it slows the heart rate. This phenomenon is known as the diving reflex and, although not very well developed in man, it is important in diving mammals like seals, as we shall see later. You can try to demonstrate the diving reflex for yourself by getting a friend to compare your normal pulse rate with that measured when you dunk your face in a basin of cold water. This experiment does not always work, however, because nervousness (or excitement) releases the hormone adrenaline, which speeds up the heart rate.

Ama: the Fisherwomen of Japan

The most famous of all skin-divers are the Ama of Japan, who harvest the submarine gardens of the ocean floor, collecting shellfish, sea-slugs, octopus, sea-urchins and seaweed. Although not commonly eaten in the western world, these are traditional delicacies in Japan. They also gather the mother-of-pearl shells known as Akoya-gai, that are used for pearl cultivation. The Ama have existed for more than 2000 years. By tradition, they are all women and they have been immortalized in the woodblock prints of the Ukioy-e artists, which depict beautiful young girls, naked from the waist up, diving for the

highly prized Awabi (abalone). The pictures are deceptive, however, for the Ama continue to dive until they reach their fifties. Nor is their job a pleasant one. Sei Shonagon, a lady-in-waiting at the court of the Japanese Empress Sadako, about a thousand years ago, described it thus:

‘The sea is a frightening thing at the best of times. How much more terrifying it must be for those poor women divers who have to plunge into its depths for their livelihood. One wonders what would happen if the cord around their waist were to break. After the woman has been lowered into the water, the men sit

Girls watching awabi-divers at Enoshima, from a triptych painted by the great Ukioy-e artist Utamaro about 1789.

Japanese Ama photographed by the Italian Fosco Maraini near the island of Hekura on the west coast of Japan. Attached to each girl's waist are a lead weight and a long rope that is used by her assistant to pull her up at the end of the dive. The knife stuck through her waist belt is used to prise abalone off the rocks.

comfortably in their boats heartily singing songs as they keep an eye on the mulberry cord that floats on the surface. It is an amazing sight for they do not show the slightest concern about the risks the woman is taking. When she finally comes up, she gives a tug on her cord and the men haul her out of the water with a speed I can well understand. Soon she is clinging to the side of the boat, her breath coming in painful gasps. The sight is enough to make even an outsider feel the brine dripping. I can hardly imagine that this is a job that anyone would covet.'

Her words sound surprisingly modern, considering the time and distance they have travelled.

Once, there were many thousands of Ama in Japan – a census taken in 1921 recorded as many as 13,000 – but their number has declined sharply in recent years. By 1963 it had dropped to 6000 and today there are probably fewer than 1000 left. Most of the current Ama are old, for few young people wish to take up such an exacting profession and many shellfish can now be cultivated artificially, obviating the need for divers. It seems likely that the Ama's profession will soon vanish, the ghost of their presence remaining only in the local village names (such as Ama-machi).

Traditionally, there are two kinds of Ama: the *cachido* and the *funado*. The *cachido* are young girls, still in training, who dive without assistance to a depth of 5-7 metres, and spend around fifteen seconds on the bottom. Although she may make as many as sixty dives an hour, the *cachido* does not run the risk of the bends because of the shallow depth. The most experienced and skilled of divers are the *funado* who dive to much greater depths – around 20 metres is average. As Sei Shonagon describes, each *funado* is assisted by a boatman. After hyperventilating to fill her lungs with air, she dives vertically to the bottom of the sea, clutching a heavy weight to help her sink and keeping her legs clamped together to reduce her resistance through the water. Once at the bottom, she releases the weight and collects her harvest, placing it in a small net basket. When she is ready to ascend, she signals to her partner by tugging on the rope attached to the weight, and he then pulls her up using the lifeline secured around her waist. Altogether, each dive lasts around a minute, about half of it being spent on the bottom. Before diving again, the *funado* rests in the water at the side of the boat for about a minute. She will usually make fifty dives each morning and a further fifty in the afternoon, but like the *cachido*, she must stop to warm up after a series of dives.

The Ama do not appear to suffer from decompression sickness, but they do experience a greater number of ear problems than their non-diving companions. A 1965 survey showed that as many as 60 per cent of *funado* over the age of fifty suffered from hearing loss. Tinnitus and rupture of the eardrum were also common complaints.

There are physiological reasons why women make better divers – they can hold their breath longer and are more resistant to cold – but somehow it seems unlikely that this is the real reason that the Ama are all women.

When you emerge from the sea, your body is no longer supported by the water and the blood redistributes from the chest to the legs. This has important implications. It was known for many years that people plucked out of the sea by an air–sea rescue helicopter were at risk of post-rescue collapse; although alive and seemingly undistressed while in the water, they suffered a cardiac arrest soon after being winched up into the helicopter. A knowledge of human physiology helped solve the problem. It was realized that the redistribution of blood on immersion in water reduced the blood flowing to the lower limbs, and allowed them to cool to a much lower temperature than that of the body core. Until a few years ago, most people were rescued in an upright position, using a winchbelt placed around the chest under their arms. Consequently, when the victim emerged from the water there was an immediate surge of blood to the legs, where it was rapidly cooled and on returning to the heart thereby induced a cardiac arrest. The solution was to place a second belt under the victim's legs which allowed them to be lifted horizontally from the water, thus preventing the blood from redistributing; and to keep them supine until the extremities had gradually rewarmed. Since this procedure was adopted by the British naval air–sea rescue services, the incidence of cardiac arrest has fallen dramatically.

Imploding and Exploding Organs

The human body is made mostly of water, which is virtually incompressible. Therefore the body remains at the same pressure as the surrounding water, and is not crushed at depth. The same cannot be said for the gases trapped in the body cavities (lungs, ears, sinuses) which, because they are compressible, occupy a smaller volume at high pressure. Shrinkage of the air in the body cavities has a number of effects, almost all of them unpleasant.

The volume of the air in the lungs of a skin-diver diminishes with depth, because of the increasing ambient pressure. It was initially assumed that this must set a limit on the depth a free diver could reach; it was argued that at some point, probably around 100 metres, the pressure would simply collapse the chest, in the same way that an empty (sealed) can, or a submarine, is crushed at depth. An alternative theory was that the rib-cage would remain intact, but that the lungs would shrink, ripping apart the delicate pleural membranes that attach them to the chest wall. Ignoring the warnings of the physiologists, some divers ventured deeper and found that they suffered no ill effect. It seems that, in this respect, humans may be more like whales and dolphins than had been expected.

There are many stories of fugitives who escaped detection by submerging themselves in a river or lake and breathing through a hollow reed. Consideration of the problem suggests that they were lucky not to have been discovered, for they cannot have been submerged very deeply. It is simply not physically possible to breathe atmospheric air when there is more than a metre of water above your head; most people cannot even manage at half a metre. This is because the external water pressure on the chest makes it harder to breathe in. Furthermore, the air in the breathing tube must also be exchanged and although reducing the diameter of the tube reduces the amount of air, it also increases the resistance, as you will find if you compare the difficulty of breathing through a straw and a snorkel tube in the swimming pool. Snorkel tubes are rarely immersed to any depth in water – they poke up above the surface and their primary function is to enable the swimmer to move along face down in the water.

At depths of more than half a metre, therefore, the diver must be supplied with air at a pressure equal to that of the surrounding water. Even then, he may not function as efficiently as on land for the density (mass per unit volume) of a gas also rises with depth, which increases the work of breathing. One solution is to substitute a less dense inert gas, such as helium, for the nitrogen in the inspired air.

The lungs are not the only body cavity that is filled with air. One of the most obvious effects of diving, with which many people are familiar, is a feeling of pressure or pain in the ears. This occurs because the air contained within the middle ear does not communicate freely with that outside. Consequently, as the air contracts on descent, pressure is placed on the eardrum, which bows inwards. To prevent it rupturing, the pressure in the middle ear must be equalized with that of the outer ear – in other words, with the external water pressure. This is achieved by admitting air via the Eustachian tube, a passage that connects the throat to the middle ear. The Eustachian tube is normally closed and usually it requires some positive action to open it. The most common way is to pinch the nose between the thumb and forefinger and try to blow your nose. Yawning also helps. Success is heralded by

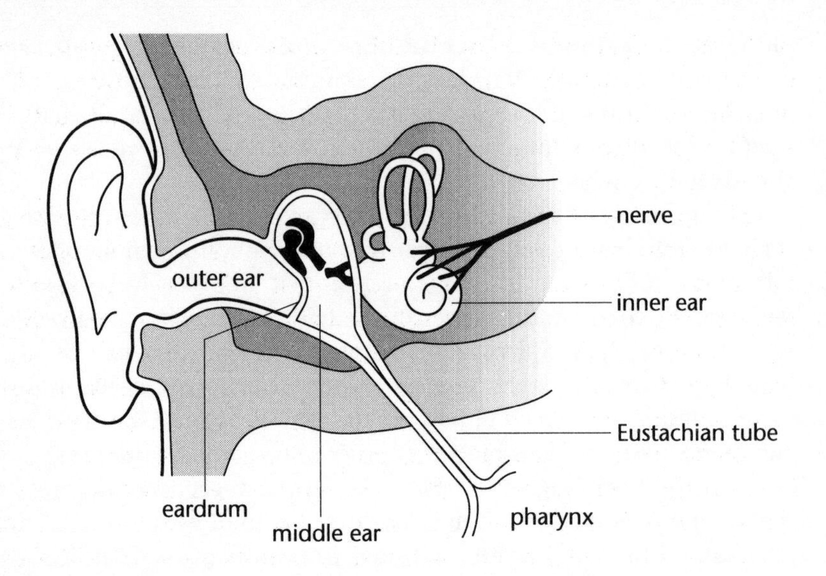

The middle ear is an air-filled cavity, surrounded by bone, that connects the outer and inner ear. It is separated from the outer ear by the eardrum and from the fluid-filled inner ear by the oval window. Sound is simply pressure waves in the air which cause the eardrum to vibrate. This is transmitted to the inner ear, where sound is detected, by three small bones imaginatively called the hammer, anvil and stapes (the latter because it has the shape of a stirrup). Divers may experience ear-pain as they sink because of the expansion of the air in the middle ear. Fortunately, the cavity is not completely sealed. The Eustachian tube, named after the Italian anatomist who discovered it, connects the middle ear to air passages behind the nose and acts as a conduit for equalizing the air pressure in the middle ear with the external air pressure.

the 'popping' of the ears as air rushes into the middle ear. It may be difficult to equalize the pressure if the Eustachian tubes are blocked by catarrh, which is why you are advised not to dive if you have a cold; it is also why flying can be uncomfortable, for commercial aircraft are pressurized to an altitude of around 2000 metres. If compression occurs so fast that the individual is unable to equalize the pressure in the middle ear, it can have unfortunate consequences. During the course of recompressing a diver suffering from severe bends, the chamber pressure was increased to 6 atmospheres within three and a half minutes, which promptly ruptured both eardrums of the doctor attending him.

A particularly nasty surprise for the diver may occur if an air bubble is trapped in a dental filling, or in a decaying tooth, as the contraction of the gas at depth may cause the filling or the tooth to implode. The converse can occur at very high altitudes, where the low pressure results in explosion of the tooth. To avoid such a calamity, Judy Leden had all her fillings replaced during her preparations for the high-altitude hang-glider flight record.

The expansion of gases when the pressure is released may also be a problem. Fish that live at great depths evert their guts when brought to the surface because the gas contained in their swim bladder expands, forcing their viscera out through their mouth. An unwary scuba-diver may also experience difficulties on ascent. At a depth of 10 metres, the pressure is twice that at the surface and air that is breathed in at this depth will therefore expand to twice its volume on reaching the surface. Ascent to the surface with a full lung of air will therefore burst the lungs. Rupture of the air-sacs allows gas to escape into the pleural cavity surrounding the lungs, or into the circulation where, because it prefers to move upwards, it may block the blood flow to the brain. This can be fatal. The ability of the lungs to accommodate expanding air is very limited and a burst lung may result from a dive only 2 metres deep. Such 'barotrauma' is exceptional, however, for providing the diver breathes in and out normally as they ascend, the volume of air in the lungs is gradually adjusted. But should it be necessary to make an emergency ascent, it is essential to remember to breathe out continuously on the way up.

Holding Your Breath

A skin-diver must overcome two main difficulties: sinking and breathing. The current record for a free (unassisted) dive on a single breath of air is 72 metres (236 feet) and was set by Umberto Pelizzari of Italy in 1992. Greater depths have been attained by 'no-limits' divers who use heavy weights to assist their descent and compressed air to jet propel themselves back to the surface. With these aids, Pelizzari descended 118 metres in 1991, but this was later surpassed by the Cuban Francisco Ferreras, who reached a staggering depth of 133 metres.

The human body is naturally buoyant as it has a density close to that of water. To dive, one must swim actively downwards, or add weights. There is a positive feedback between depth and buoyancy, due to the role played by the air in the lungs: the deeper a breath-holding diver goes, the denser he becomes, because the air in his lungs is compressed and provides less buoyancy. Thus, he sinks more rapidly. Conversely, the higher the diver rises, the more the air in the lungs expands, making him lighter and causing him to ascend faster. This means that although it may take quite an effort to get down the first few feet, it becomes gradually easier, until at around 7 metres the diver will naturally sink. It therefore becomes increasingly difficult to swim up from deeper waters, which is why most deep divers (such as the Japanese shellfishers) are hauled up by an assistant.

Buoyancy

Animals use many wonderful devices to maintain their vertical position in the ocean. Most avoid expending unnecessary energy by ensuring that their density matches that of the surrounding water. This is the function of the swimbladder, a silvery, air-filled sac that can be observed lying within the body cavity of a fish when it is gutted. This organ enables the fish to adjust its buoyancy to the depth at which it lives. Neutral buoyancy is beneficial, as the fish need not expend energy to maintain its horizontal position, but it also carries an inherent disadvantage. Like the human skin-diver with a lung full of air, if the fish swims below its usual depth, the gas in the swimbladder is compressed and it must swim harder to prevent itself from sinking. Conversely, if the fish swims above its neutral buoyancy depth, the gas expands and provides additional lift, so that the fish must now swim downwards to avoid being carried to the surface. Although a fish can adjust its neutral buoyancy by secreting or removing gas from the swimbladder, this is a slow process, so that fish are essentially confined to a slice of the ocean and, like aircraft circling in a stack above an airport, each species has its own cruising depth. Many fish have closed swimbladders with no external openings and if they are brought rapidly to the surface, the gas may expand so quickly that the swimbladder is ruptured or forced out through the mouth. Some fish (sharks, for example) do not have swimbladders at all and must swim constantly to maintain their position in the water – they sink if they stop. The basking shark, however, which as its name implies spends less time rushing around, has a large oily liver that helps it achieve neutral buoyancy.

The swimbladder is filled almost entirely by oxygen which is prevented from diffusing out by the fact that the organ is lined by multiple layers of guanine crystals. These crystalline layers may also protect the cells that form the walls of the swimbladder from the toxic effects of oxygen at depth. Guanine itself is a most interesting molecule for it gives

The biggest problem faced by skin-divers, of course, is lack of air. Most people are unable to hold their breath for more than one or two minutes, but with training longer times can be achieved. The world record is six minutes and forty-one seconds, and was set by Alejandro Ravelo in 1993, while lying quietly on the bottom of a swimming pool. To attain such records it is necessary to hyperventilate before diving. As explained in Chapter One, carbon dioxide provides the main stimulus for breathing, so hyperventilation, which blows off additional carbon

fishscales their shine, is found in bird excrement (it is the main constituent of guano), and, most importantly of all, is one of the four bases that make up DNA.

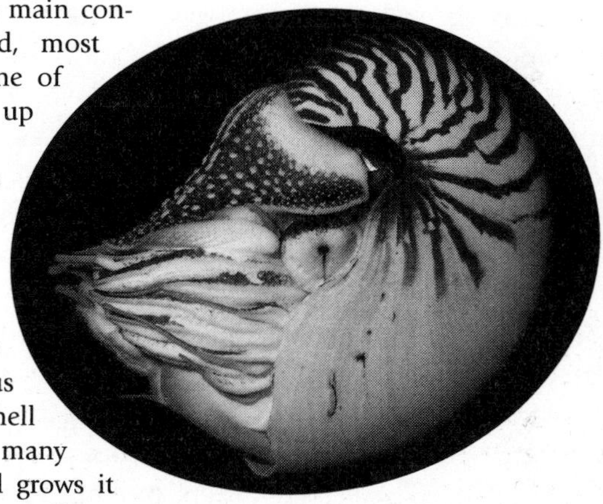

The Pearly Nautilus is a beautiful creature related to the ammonites of ancient time, and to the octopus and squid of today. It is also known as the Chambered Nautilus for it has an external shell that is divided into many chambers. As the animal grows it adds new chambers to its shell, at a rate of one every three to four months. Each chamber is separated from its neighbour by walls known as septa which strengthen the shell and help prevent it from being crushed by the external water pressure. The animal lives in the last chamber; the others are filled with gas at atmospheric pressure and used for buoyancy. When first formed, the chamber is filled with salt solution, but the salts are gradually pumped out, dragging the water osmotically with them, which allows gas to diffuse in and replace the liquid. Because the gas used for buoyancy is confined within a rigid shell, the Nautilus is unaffected by changes in depth and is free to hunt vertically within the ocean, being limited only by the pressure which its shell can support. During the day it descends to around 400 metres, but at night it rises up to shallower waters (150 metres deep) to feed. It has been captured at depths as great as 600 metres, but experiments have shown that the shell is crushed by an external water pressure of around 750 metres. This, then, is the ultimate limit for the Pearly Nautilus.

dioxide, prolongs the time before the carbon dioxide concentration rises to a level which stimulates the next breath. Hyperventilation before diving is very dangerous, however, because the diver may not be aware that the oxygen level in their blood has fallen too low for normal brain function and they may pass out underwater and drown. Even today, this causes unnecessary deaths, usually of children competing to see how long they can remain underwater.

Although humans cannot hold their breath for more than a few minutes, diving mammals, ducks and turtles can do so for far longer. The elephant seal holds the record and a single dive of two hours' duration has been timed, which is more than twenty times the human limit. The majority of dives, however, are far shorter. The enormous endurance of the seal is not because it carries more oxygen in its lungs, for, as we have seen, it actually exhales before diving to avoid the bends. Relatively speaking, however, seals and whales have a larger blood volume and a higher oxygen-carrying capacity than a human, so that the amount of oxygen carried in the blood is much greater. Oxygen is also stored in the muscles, bound to myoglobin, a molecule that is structurally similar to haemoglobin, the oxygen-carrying pigment of the blood. Sperm whales have ten times more myoglobin per kilogram of muscle than man, which accounts for the very dark red colour of whale meat. In addition, the muscles of diving mammals contain a large amount of creatine phosphate, which acts as an energy supply (see Chapter Five). These adaptations provide Weddell seals and whales with an oxygen supply that lasts for about twenty minutes – somewhat longer than the length of a normal dive.

On occasion, Weddell seals may make protracted dives of up to an hour's duration. This is possible because after consuming the oxygen stored in myoglobin, the muscle switches over to anaerobic metabolism, which does not utilize oxygen (see Chapter Five). Anaerobic metabolism, however, results in the formation of lactic acid, which must be removed from the tissues subsequently in a process which requires oxygen. Thus, the longer the seal stays submerged, the more lactic acid builds up, and the larger the amount of oxygen needed on surfacing to get rid of it. This explains why, after a lengthy dive, a Weddell seal remains at the surface for a longer time before its next dive.

The elephant seal remains an enigma. Like the Weddell seal, its oxygen stores last for only about twenty minutes. Yet it is able to remain underwater for well over an hour, and then dive again immediately after

The elephant seal, the greatest of diving mammals

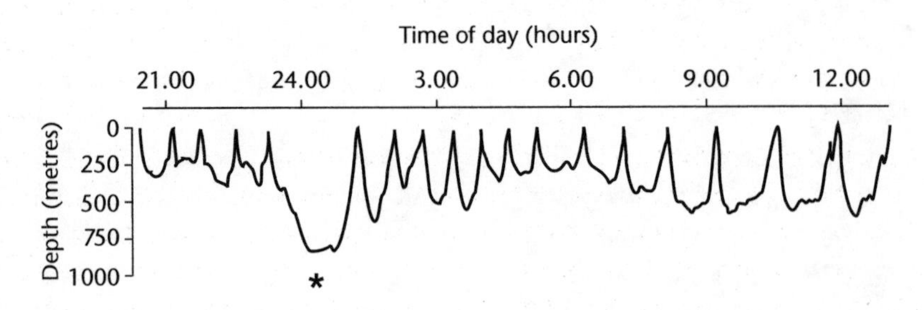

Part of the diving record of a free-ranging female elephant seal (*Mirounga angusti*) which had been tagged with a radio-transmitter. Most of the time is spent below the surface and one dive (marked with an asterisk) lasted for two hours.

surfacing. Obviously it has no lactic acid to get rid of, which means that its oxygen supply must last for far longer than predicted. No one really knows how it manages this feat, but one suggestion is that the metabolic rate drops sharply during deep dives. In many diving animals, including elephant seals, the heart rate drops instantaneously on submerging, a phenomenon known as the diving reflex. Blood vessels serving the skin and viscera constrict, channelling the blood away from these less vital organs towards the brain and heart. The metabolic rate drops in the less well perfused tissues, reducing their oxygen requirement. This redistribution of blood may therefore help conserve the limited oxygen supply. But this is mere speculation, and precisely how the elephant seal actually manages to dive for such extended periods is still unclear.

Several other mysteries also remain. Platypuses, for example, seemingly like to meditate underwater, as they often lodge themselves under tree roots and lie at the bottom of a stream for considerable periods. The green turtle *Chelonia mydas* overwinters at the bottom of the Gulf of California where it lies submerged among the mud and eel-grass in a dormant state for several months. Although the metabolic rate is much lower during dormancy, it is still not clear how the turtle obtains sufficient oxygen for its needs. Sadly, we may no longer have the opportunity to find out, for the wintering sites, once known only to the local Seri Indian population and carefully conserved, have recently been discovered by Mexican fishermen with modern equipment, who have rapidly depleted the turtles' numbers.

Scuba-diving

Diving was revolutionized in the middle of this century by the introduction of the *s*elf-*c*ontained *u*nderwater *b*reathing *a*pparatus (SCUBA). The key was the development of the demand valve, in 1943, by two Frenchmen, Jacques Cousteau and Emile Gagnan. As its name suggests, this device supplies air to the diver as required, at the same pressure as the surrounding water. The rest of the scuba equipment consists of one or more tanks of compressed air, carried on the back, a face-mask and fins. Parenthetically, it seems surprising that the fin was not introduced until 1935 and even then only in crude form (a wood and metal paddle), because it has a remarkable effect on a swimmer's efficiency.

Scuba-diving was first used to locate and remove enemy mines after the Second World War, but in the 1960s it was introduced to the general public in a series of wonderful underwater films taken by Cousteau and by a German couple, Lotte and Hans Hass. Their films of coral reefs, dolphins, sharks, and many more unusual sea-creatures revealed the variety and complexity of life in the oceans. People were captivated by a psychedelic world in which man appeared able to fly effortlessly among myriad clouds of brilliantly coloured fishes; where animals were inquisitive rather than fearful; where treasure lay scattered on the sea-bed for all to find; and where few had ever explored before. Many desired to it see for themselves, fuelling the development of a scuba-diving industry that now supports a large number of recreational divers. Yet, as we have already seen, for all its beauty, the underwater world is not without its dangers, and would-be scuba-divers are well advised to undertake a reputable training course before they venture beneath the surface.

The safe limit for compressed-air diving, whether as a scuba-diver or with a surface air supply, is around 30 metres. This floor is imposed by the gases in the air you breathe, for, at pressure, both nitrogen and oxygen act as poisons.

The Rapture of the Deep

At a pressure of several atmospheres, nitrogen has an intoxicating effect that has been christened 'the rapture of the deep' by Jacques Cousteau.

The symptoms take some time to develop and are similar to those produced by alcohol: elation, perceived mental agility, detachment from reality, loss of manual dexterity and irrational behaviour. The sense of euphoria is both illusory and dangerous, for if the diver continues to descend he will become more and more confident but less and less capable. Mild nitrogen intoxication occurs at a depth of about 50 metres. At greater depths the symptoms become more marked until finally consciousness is lost, usually around 90 metres. With frequent exposure, the diver may become somewhat habituated to the effects of nitrogen, a phenomenon known as 'working up', which enables him to venture to 50 metres without becoming severely intoxicated. Nevertheless, nitrogen narcosis has been responsible for the death of many divers attempting deep dives and it accounts for the advisory diving limit, when using compressed air, of 30 metres.

The scientist J.B.S. Haldane – the son of J.S. Haldane – made a scientific study of the effects of nitrogen intoxication using a pressure chamber in 1941. His subjects, who included himself and his future wife, were set arithmetic and manual dexterity tests, the latter consisting of transferring small ball-bearings from one jar to another using forceps. When breathing air at a pressure of 10 atmospheres (equivalent to a depth of 90 metres), all of them became rather confused. One individual, reported to be a thoroughly responsible scientist at atmospheric pressure, cheated in the dexterity test and another alternated between depression and elation, at one moment begging to be decompressed because he felt 'b..... awful' and the next minute laughing and attempting to interfere with his colleague's dexterity test. No one managed to get their sums right – as Haldane laconically noted, 'observations were not as satisfactory as could have been wished.' Another difficulty was that the person administering the test was usually as intoxicated as their subject, and often failed to take proper notes or to stop the stopwatch. Such studies sufficed to show that divers suffering from nitrogen narcosis could not be expected to behave responsibly and might react in ways that threatened their own or others' lives. Indeed, some intoxicated scuba-divers have been known to offer their mouthpieces to the passing fish.

Recovery from nitrogen poisoning occurs remarkably quickly on decompression. In Haldane's experiments, an immediate relief of symptoms was produced when the pressure was reduced from 10 to 5 atmospheres. The typical reaction was, 'My God, I'm sober.'

Why should nitrogen at pressure cause narcosis? This question has still not been answered. The similarity of the symptoms suggests that the mechanism may be the same as that involved in the action of alcohol, but this does not help much since we still know little about how alcohol works. The most recent studies suggest that alcohol interacts with a particular class of proteins in cell membranes, known as ion channels, that regulate the excitability of nerve cells. Perhaps, then, this may be how nitrogen acts.

Too Much of a Good Thing

Oxygen is a toxic substance and it becomes increasingly so as the pressure is increased.[5] Although most people can safely breathe pure oxygen at 1 atmosphere for up to twelve hours without harmful effects, after about twenty-four hours they begin to develop lung irritation, due to the progressive destruction of the cells which line the air-sacs. The first sign of a problem is coughing, but in severe cases this may progress to difficulties in breathing, leakage of fluid into the lungs and even haemorrhage of the lung capillaries, so that the lungs fill with blood. At 2 atmospheres pressure, the nervous system is also affected and the individual may suffer from vertigo, nausea and paralysis of the arms and legs. Convulsions, resembling a major epileptic fit, set in within a few hours and even more quickly with physical exertion. Sometimes, these are violent enough to break bones. As the atmospheric pressure is increased, the time before convulsions occur decreases. Any convulsion underwater is obviously potentially fatal and must be avoided, so extensive experiments were undertaken, again by J.B.S. Haldane, during the Second World War. He observed: 'The convulsions are very violent and in my own case the injury caused by them to my back is still painful after a year. They last for about two minutes and are followed by flaccidity. I wake in a state of extreme terror in which I may make futile attempts to escape from the steel chamber.'

Haldane and his colleagues found that at a pressure of 7 atmospheres exposure to pure oxygen was limited to about five minutes before convulsing. To his delight, Haldane also discovered that, at this pressure, oxygen was not the odourless and tasteless gas it appeared to be at atmospheric pressure. Instead it had a rather peculiar flavour, being both sweet and sour, 'like flat ginger beer' or 'dilute ink with a little sugar in it'. He liked to use this fact to illustrate that one should not believe everything that one reads in textbooks, which invariably state that oxygen is tasteless.

During the Second World War, the British navy was using (and still does) a closed-circuit rebreathing apparatus supplied with pure oxygen. This consists of a counterlung, which is carried on the chest, and an oxygen cylinder. The counterlung is a large flexible rubber bag that expands and collapses as the diver breathes. Between the mouth and the counterlung lies a carbon dioxide scrubber (filled with soda-lime) which removes the carbon dioxide exhaled by the diver. Oxygen is fed into the counterlung to replace that used by the diver. No gas is released into water, and thus no tell-tale bubbles are formed. This is a great advantage in clandestine operations where the diver must remain undetected. It is also useful when defusing underwater mines by hand, because bubbles may set off the mine. A further advantage is that the gas cylinder can be one-fifth the size of a SCUBA tank (because air only contains 20 per cent oxygen) which makes the diver more manoeuvrable;[6] or alternatively, if he carries the larger tank, extends his range. As a result of Haldane's experiments, the limit on diving when breathing pure oxygen was set at a depth of 8 metres (1.8 bar). Even then, this is only tolerable for a few hours. Some people are more susceptible to oxygen toxicity than others, so the British navy currently test their new diving recruits by pressurizing them to 2 atmospheres to see whether they have a fit when given pure oxygen to breathe. Those who do so are offered a different specialist training.

Below 8 metres, pure oxygen cannot be used and it becomes necessary to feed the counterlung with a mixture of gases. At depths down to about 25 metres this is usually 60 per cent oxygen and 40 per cent air, and the percentage of oxygen is further reduced at lower depths, falling to 33 per cent at 50 metres. The disadvantage of this gas mixture is that the nitrogen in the air accumulates in the counterlung, making it necessary to flush the system at intervals. Although this creates bubbles, it only does so periodically, so the counterlung remains the system of choice for covert operations like planting bombs on enemy ships. The decompression time is also much lower because little nitrogen is present in the gas mixture.

Toxicity is also an important consideration when breathing lower oxygen concentrations, similar to those found in air, at depth. As the diver descends, the pressure of the air she breathes increases in parallel with the water pressure. At a depth of 90 metres, for example, the pressure is 10 atmospheres. Since one-fifth of air is oxygen this means that the partial pressure of oxygen is now 2 atmospheres. Although this can be tolerated for a short time, it is not desirable for long dives and the oxygen content of the inspired gas must be decreased. Deep-diving animals like whales and seals do not suffer from oxygen toxicity, or nitrogen narcosis, as they do not breathe air under pressure – indeed, no air leaves their lungs during a dive.

Blackouts and Bravery

We should also consider the effects of carbon dioxide at pressure which, while perhaps not as dramatic as those of nitrogen and oxygen, may also have serious effects. As described in Chapter One, carbon dioxide acts as a potent stimulus to breathing. A rise in carbon dioxide not only increases the respiratory rate, however; if it continues, it can also lead to headache, confusion and loss of consciousness.

In the early part of the twentieth century, carbon dioxide poisoning was found to be the explanation for why so many British naval divers were unable to do any work at depth. The diver was constantly supplied with air from the surface, which escaped through a bleed valve on the side of his helmet. Carbon dioxide is a waste product of metabolism that is excreted in the exhaled air. Respiration, therefore, raised the percentage of carbon dioxide in a diver's suit to a level above that in the air intake, by an amount that depended on the rate at which the air flowed through the suit. Exercise, which increases the metabolic rate, further increased the concentration of the gas. A carbon dioxide level of 2 per cent had little effect on a diver's performance at normal atmospheric pressure, so the rate at which air was supplied to him was calculated to ensure that this limit was not exceeded. At that time, however, it was not appreciated that the effects of carbon dioxide are exacerbated by pressure and that at a depth of 60 metres, where the pressure is 5 atmospheres, 2 per cent carbon dioxide has an effect similar to that produced by a 10 per cent concentration of the gas at the surface. Consequently, if the diver tried to exert himself, not only did he pant excessively but he also often lost consciousness. Once the cause of the problem had been identified, it was easily rectified by increasing the rate at which the air was supplied in proportion to the external water pressure.

J.B.S. Haldane (1892–1964) was a brilliant and influential British scientist. His studies on the effects of gases on the human body at pressure transformed diving practice, but his greatest work was as a geneticist and on the mathematical basis of evolutionary theory. A flamboyant, argumentative and controversial character, he was also a dedicated Marxist and a highly successful popularizer of science, regularly writing scientific articles for the *Daily Worker.*

Carbon dioxide toxicity may also occur when using the closed-circuit rebreathing apparatus described above, if the soda-lime scrubber that removes the gas is not functioning properly or becomes exhausted. This may be one reason why a number of naval divers during the Second World War blacked out underwater and consequently drowned, even though they were operating at shallow depths.

Further studies into the effects of breathing carbon dioxide under pressure were initiated as a result of tragedy. Three months before the start of the Second World War, in June 1939, the British submarine *Thetis* sank off Liverpool during sea trials, with the loss of ninety-nine lives. Only four men survived. Once more, J.B.S. Haldane was called in, this time by the trade unions to which many of the men belonged, to investigate the cause of their death. He employed four non-scientific assistants, all volunteers.[7] To simulate the conditions in the escape chamber of the submarine, they were interned in a small steel chamber. Within an hour they all had splitting headaches, and several of them vomited, due to the increased carbon dioxide concentration.

About 3 per cent of expired air is carbon dioxide, so that if people are confined in a small space and are forced to rebreathe the same air, the carbon dioxide level rises. In a stricken submarine, a rise in carbon dioxide may occur before it is appreciated that it is necessary to abandon ship: in the case of the *Thetis*, it appears to have risen to around 6 per cent (its normal value in the atmosphere is 0.04 per cent). But this is not the only problem, because the partial pressure of carbon dioxide in the air rises even higher when the escape chambers are used. In a submarine, the escape hatches open outwards, so that the external water pressure helps seal them shut. In order to open them, the air pressure inside the submarine must be equalized with that of the outside water, by flooding the escape compartment with sea water. As soon as the pressure gradient is dissipated, and the escape hatches can be opened, the crew put on their breathing equipment and ascend to the surface. Because the air in the escape chamber is compressed as the water enters, the partial pressure of carbon dioxide will gradually go up.

Subsequently, therefore, Haldane, together with Dr Martin Case, carried out extensive tests on the effect of raising the carbon dioxide concentration under increased pressure. A rise from 0.04 per cent to 6 per cent had little effect at 1 atmosphere, but at 10 atmospheres there was a marked deterioration in performance on dexterity tests, all of the subjects became confused and most of them passed out within five minutes. Underwater, confusion or unconsciousness can have fatal consequences. Haldane's studies therefore suggest that when the escape chamber in the *Thetis* was suddenly decompressed, the carbon dioxide concentration in the remaining air may have been high enough to impair the submariners' judgement and hinder them in putting on and correctly adjusting their breathing apparatus.

As must be evident by now, J.B.S. Haldane was quite an eccentric and delighted in testing his own body (and those of his colleagues) to the extremes. He was also a very thorough scientist, so he subsequently investigated the effects of carbon dioxide at the cold temperatures encountered at depth. He wrote that on one occasion he was

> 'immersed in melting ice for thirty-five minutes, breathing air containing 6.5 per cent carbon dioxide and during the latter part of the period also under ten atmospheres of pressure. I became unconscious. One of our subjects has burst a lung but is recovering; six have been unconscious on one or more occasions; one has had convulsions.'

One wonders what the Health and Safety Executive would have to say about such studies today. Yet Haldane's personal bravery, and that of his team, provided the data necessary for a scientific understanding of the effect of gases on the human body under pressure. The knowledge they gained saved many lives and continues to do so today.

How Far Can You Go?

The danger of nitrogen narcosis means that compressed air cannot be used at depths greater than 30 metres. The nitrogen must be replaced and, as the the diver descends, the amount of oxygen must be continuously adjusted to ensure the pressure never exceeds 0.5 bar. The balance of the inspired gas is made up with helium, and at depths below 30 metres divers usually breathe a helium–oxygen mixture known as heliox. Helium has several advantages over nitrogen as an inert gas. First, it produces much less narcosis. Secondly, it is easier to breathe, as it is less dense and thus less viscous: its molecular mass is only 4 compared with 28 for nitrogen. Helium is also significantly less soluble in water than nitrogen, which reduces the amount of gas that dissolves in the blood and thus lowers the time required for decompression. On the negative side, helium has high thermal conductivity, which means that a lot of heat is lost through the expired air, so the diver may need an additional heating system to stay warm. And because of its low density, the pitch of the voice is increased, giving a squeaky intonation like a cartoon character. This 'Donald Duck' voice results because the vocal cords vibrate more rapidly in the lighter air.

At depths of over 200 metres (21 bar), humans and other land-living animals develop high pressure nervous syndrome (HPNS). This is a neurological disorder, colloquially known as 'the shakes' because it causes tremor. Dizziness, nausea and short sleep-like periods of inattentiveness are also symptoms. The cause of HPNS is not well understood but it may result from a direct effect of pressure on the nervous system, as isolated nerve cells show a similar hyperexcitability when pressurized to an equivalent depth in the laboratory. Remarkably, the effects of pressure and anaesthesia interact. Whereas tadpoles cease swimming if they are exposed either to a low concentration of alcohol (2.5 per cent) or to high pressure (20–30 bar), when both are applied together they swim around happily. Likewise, mice given a general anaesthetic wake up if the pressure is increased while, conversely, HPNS is reduced by general anaesthetics. This experiment has never been tried directly on humans, but the animal experiments led to the discovery that HPNS can be partially overcome by the addition of a small amount of nitrogen to the heliox mixture. This is known as trimix gas.

HPNS limits the depth to which a diver can descend without an artificial environment. When breathing heliox, the limit is 200–250 metres, but experimental dives suggest that humans can operate down to depths as great as 450 metres in the open sea (and 600 metres in a pressure chamber), provided that they breathe special gas mixtures, such as trimix. Nevertheless, these regions remain the prerogative of the test pilots of the deep and are not normally visited by man. In contrast, depths of over 200 metres are routinely visited by marine mammals: sperm whales may dive to 1100 metres and elephant seals have reached depths of 1500 metres. Many other types of animal – fish, bacteria and polychaete worms – live at even greater depths around the mid-ocean vents. So why don't they get high pressure nervous syndrome? Studies of deep-sea species reveal that these animals exhibit much higher pressure thresholds for HPNS. Moreover, they seem to require high pressure for normal physiological function since, in their case, *decompression* may lead to symptoms resembling those of HPNS. They may therefore be considered 'obligate barophiles'. Currently, scientists are trying to unravel the puzzle of how their cells are able to function at such extreme pressure.

Living at Depth

As we have seen, more gas dissolves in the body fluids at depth because of the increased pressure. At extreme depths, the decompression time, even for a very short dive, may be many hours, so that it becomes impractical to return directly to the surface. Instead, the diver both lives

Underwater Challenges

Pressure is not the only problem for the diver. The intense cold at depth, and the fact that they are weightless in water contribute to their difficulties. Vision, hearing and orientation are also affected.

Almost all divers wear goggles or a face-mask as without them the eyes are unable to focus underwater and everything appears blurred. This is because when a light ray passes from one medium to another – in this case from air (or water) into the eye – it is bent (refracted). This property is used to help focus the light rays on the layer of light-sensitive cells, known as the retina, at the back of the eye. The extent to which a light ray is bent at the surface of the eye is very much less in water than in air, which makes it impossible to focus the image on the retina. Maintaining an air space next to the eye, by wearing goggles or a face-mask, obviates the problem. But because the light rays will now be refracted by the glass/water interface of the mask, objects appear some 30 per cent larger and closer underwater than they do in air. It may be useful to remember this when listening to divers' tales of giant sharks.

Water absorbs light, so that the light intensity decreases with depth, and by 600 metres down the ocean is completely dark. Since red light is absorbed more easily than blue light, water also acts as a colour filter. With increasing depth, first the reds and yellows, and then the greens, are rinsed out until finally only blue remains. William Beebe described this colour change in the words of a poet. Fifteen metres down in his bathysphere he looked out at a 'brilliant bluish-green haze' which was slowly transformed by 'a slight twilighting and chilling of the green' as he descended, until at 100 metres it was a pure pale blue. At a depth of around 200 metres the light was an 'indefinable translucent blue quite unlike anything I have ever seen in the upper world and it excited our optic nerves in a most confusing manner'. The brilliance of the blue was enhanced by the searchlight, which 'seemed the yellowest thing I have ever seen'. Slowly, as he descended further, his achingly beautiful blue faded to a dark inky colour, but it had already made a lasting impression on him. Other explorers report that the blue light turns an intense violet before finally being replaced by a velvet blackness, darker than night.

A fascinating aside is that Beebe's account was clearly read by Thomas Mann, for he incorporated it into his novel *Dr Faustus*. Adrian pretends to have set a new deep-sea record with an American scholar Akercocke. He relates how 'he and Professor Akercocke climbed into a bullet-shaped diving bell of only one point two metres inside diameter, equipped somewhat like a stratosphere balloon, and were dropped by a crane from the companion ship into the sea, at this point very deep' They 'plunged into the water, crystal-

William Beebe (left) and Otis Barton (right) next to the bathysphere in which they made their epic journey 'half a mile down'. Beebe was a well-known naturalist and an author of many popular science books. Barton was an adventurous and wealthy young man with a passion for exploration, who designed the bathysphere and paid for its construction. It had steel walls nearly 4 centimetres thick and was tethered to the mother ship by a 1067-metre steel cable. The entrance was a circle only 35 centimetres in diameter so that the occupants had to squeeze in and out head first. The windows were made of fused quartz, 7.5 centimetres thick. Inside, the life-support system included oxygen tanks, and trays of calcium chloride (to absorb water vapour) and soda lime (to absorb carbon dioxide). During their descent into the deep, Beebe and Barton not only observed living specimens of fish previously known only from dead carcasses fished up in nets, but also unknown luminescent creatures. Beebe remarked that he felt 'like a palaeontologist who could suddenly annihilate time and see his fossils alive'.

clear at first, lighted by the sun'. But this illumination 'reached down only some fifty-seven metres' and at lower depths 'through the quartz windows the travellers looked into a blue-blackness hard to describe'. After that came 'solid blackness all around, the blackness of interstellar space whither for centuries no weakest sun-ray had penetrated'.

The colour of an object is determined by the wavelength of the light that it reflects: a red rose, for example, appears red because it reflects red light and absorbs all other wavelengths. Twenty metres down in the Mediterranean sea, the same red rose would appear black because there is no longer any red light to reflect. At greater depths, the light intensity is so low that the colour-sensitive cells (the cones) in the retina of the eye are unable to function. Everything then appears grey. When it is very dark, as at twilight and in the ocean depths, we use a different set of retinal cells to detect light. These are the rods, which cannot discriminate colour but are very sensitive to light – so sensitive, in fact, that they are inactivated by bright daylight and take twenty to thirty minutes to recover when the light is dimmed. Anyone who has sat for some time in a darkened room and noticed how the mysterious shadows slowly resolve into recognizable objects will be aware of this. Most divers do not spend enough time at depth to adapt fully to the dark. However, because the rods are quite insensitive to red light, a removable red visor fitted to the outside of the face-mask and worn prior to diving (and removed at depth) might help improve their vision.

Part of the attraction of diving, gleaned from films or personal experience, is the silence of the underwater world. It is much more difficult to hear underwater than in air, for sound attenuates more rapidly in the denser medium. In addition, because the soundwaves travel more quickly through water, they tend to arrive at both ears at about the same time, making it hard to localize where a sound is coming from.

The oceans are too cold for man to survive in them for very long without insulation (the surface water of tropical seas are an exception). Cold water conducts heat away from the body very effectively, so some additional form of thermal insulation for the diver is essential. *Wetsuits* work by trapping a thin layer of water between the body and the latex of the suit, while *drysuits* exclude the water and are usually worn over several layers of clothing. At depths of more than 50 metres, heat loss is exacerbated by the need to breathe heliox gas. Helium has a high thermal conductivity so that a lot of body heat is lost in breathing. It is therefore often necessary to provide deep-sea divers with their own personal heating system by piping hot water through the diving suit and, in some cases, even by warming the gas supply.

Divers are essentially weightless because of the buoyancy of the water. This liberation from the restraints of gravity is one of the great joys of diving but it is not without its difficulties. In particular, it is hard to use tools that require torque, for your whole body spins when you apply force to a spanner while the nut that you are trying to undo remains firmly locked in place. It also makes it difficult to stay in the same spot when a current is running. At great depths, the in-creased density of the water also magnifies the effort required to make a movement and limits the amount of work that can be undertaken.

On land, gravity and visual cues tell us about the position of our body. This information is not available to a weightless diver in poor visibility and may produce disorientation and alarm. It is easy to panic when it is not immediately clear which way the surface lies. Fortunately, there are still clues: bubbles always rise upwards, a released weightbelt always falls downwards.

and works at depth, returning after his shift to a living capsule that is maintained at the same pressure as the surrounding water. This is known as *saturation diving*, because the duration of the dive means that body tissues become fully saturated with nitrogen. In recent years, saturation diving has become relatively common and men may remain at depth for several weeks before returning to the surface. A stint of one month is usual for divers on the North Sea oil rigs who are involved in constructing and repairing pipelines on the ocean floor.

Saturation divers usually breathe heliox, the exact composition of the inspired gas depending on the 'storage' depth at which they live. One of the major drawbacks of breathing heliox is its effect on speech, but an electronic device known a helium speech unscrambler can be used to counteract this effect and render the diver's speech more intelligible. Because of the high thermal conductivity of helium, which saps the body's heat, the living quarters also have to be kept at a temperature of around 30°C. Otherwise, the daily difficulties of living at pressure are few. The most obvious is the tedium of the long decompression times: four days are needed for decompression following a saturation dive to 100 metres and ten days when ascending from a depth of 300 metres. During this time, the diver can do little but sit and wait. Even when atmospheric pressure is finally reached, commercial divers are required to remain in the vicinity of a decompression chamber for several hours in case decompression sickness occurs. This is because, even when conducted according to authorized schedules, around 1 per cent of dives lead to some form of the bends, which may require treatment in a recompression chamber.

A medical emergency is a real problem if it happens in a saturation vessel, as it can take many hours to get a doctor down to the living capsule. All saturation divers therefore need to be familiar with the treatment of hyperbaric illness, and in large diving teams some individuals are trained in advanced techniques, such as setting up an intravenous drip or giving a local anaesthetic. For really serious problems, however, the diver has to be evacuated. The quickest and safest way to do so is to keep him at the storage pressure using a hyperbaric transfer chamber, like that operated by the National Hyperbaric Centre in Aberdeen, Scotland to assist divers working in the North Sea oilfields. They use a one-man chamber to bring the ill or injured diver from his living quarters at the bottom of the sea to the ocean surface. This is then placed in a helicopter and connected to a larger two-man chamber in which a diving doctor is waiting to attend the victim during the journey to dry land. On arrival, the diver is transferred, still at the same pressure, to a large medical chamber where he can be safely treated. All saturation diving vessels working in the North Sea must also have hyperbaric life-boats that can accommodate several people, in case the living capsule has to be evacuated for any reason.

Long-term Dangers

The long-term effects of working under pressure were discovered almost a hundred years ago in compressed air construction workers. Several of them reported disabling pains in their hip and shoulder joints, sometimes long after they had ceased working under pressure, and on X-ray examination their joints showed evidence of degeneration. The first case of bone damage in a diver was not described until thirty years later, but from then on a steady trickle of cases began to be reported.

By the mid-1960s, the evidence was incontrovertible. In a study of 131 German divers over a ten-year period, seventy-two were found to have bone necrosis on radiological examination and only twenty-two showed no sign of disease. Likewise, 20 per cent of caisson workers involved in the construction of the Clyde tunnels had bone lesions. The damage was mainly located at the ends of the long bones of the legs and arms, and is thought to result from the presence of tiny gas bubbles in the bone tissue which block the fine capillaries supplying the bone cells and lead to their death. One reason why bone may be particularly susceptible to these microbubbles is that if a bubble forms, the living bone cells will be squeezed to make room for it, for bone itself is indistensible. In a few people, the articular surfaces of the bones may also be affected, causing severe arthritis in the hips and shoulders.

As might be expected, the frequency and severity of bone disease is related to the diving depth – no damage is found in those who have never dived below 30 metres while around 20 per cent of those who have been below 200 metres show signs of necrosis. Today, commercial divers have regular bone scans so that they can stop diving in time to prevent bone collapse.

Divers may also suffer from long-term hearing loss. Why this should be the case is not entirely clear. One idea is that working underwater can be very noisy, for air rushes in and out of the chamber during

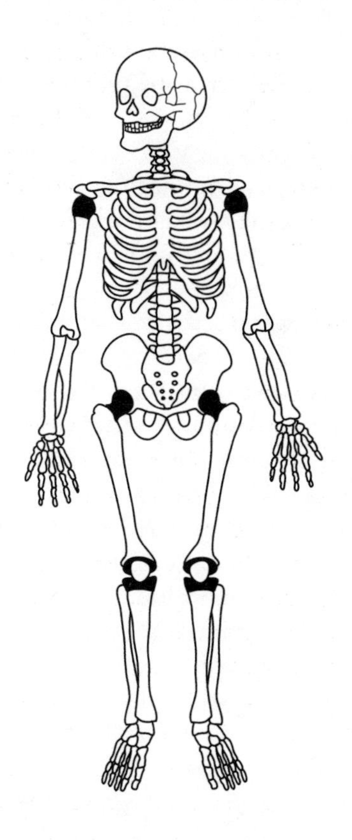

Sites of bone lesions in 72 divers found in a study of 131 divers conducted in Kiel, Germany

compression and decompression, gas continuously circulates through diving helmets and underwater construction tools can be as noisy as their land counterparts. But noise-induced hearing loss is not the only explanation. Trauma, produced by difficulties in equalizing the pressure in the ears or by fine bubbles formed on decompression, is an alternative possibility and is almost certainly the cause of the impaired hearing that affects Japanese shellfish-divers.

Many studies have examined whether diving causes brain damage. There is general agreement that divers who have suffered severe decompression sickness may sustain lasting neurological damage, but the jury is still out as to whether subclinical damage may occur in divers who have never experienced decompression problems; some studies suggest they exhibit increased tremor, reduced sensation in the feet and hands, and other signs of neurological dysfunction, while other studies show no clear effect. Because of the increasing number of recreational divers, more research is clearly needed.

In 1997, a worrying report appeared in the British Medical Journal. Using nuclear magnetic resonance imaging, tiny lesions were detected throughout the brains of some scuba-divers. These lesions correspond to areas of dead nerve cells and are thought to result from blocking of the blood supply by tiny air bubbles. Not all scuba-divers had holes in their brains. Closer examination revealed that this was only observed in people who had a small hole between the right and left sides of the heart. Surprising as it may seem, this is fairly common and is found in about a quarter of the population. It arises because during development the left and right atria – the low-pressure chambers of the heart – are connected by a hole known as the foramen ovale (literally, the oval orifice). At birth this hole normally closes, but in some people the closure is not quite complete. In these individuals, tiny bubbles formed in the circulation during decompression, too small to cause the bends, may pass through and lodge in the cerebral circulation (in other people they get trapped in the lung capillaries where they do little harm). Although there was no evidence of obvious neurological damage in this study, people with an open foramen ovale might be well advised to avoid scuba-diving.

Into the Abyss

Divers breathing heliox can descend to depths of 200 metres if they are physically fit and well-trained. If exotic gases are used, this can be extended to almost 400 metres but the diver must wear a fibreglass helmet and a heated suit. Below this depth, man must take his environment with him. Submersibles have the distinct advantage that the crew live at normal atmospheric pressure so there is no necessity for a prolonged decompression and the vessel may both sink and ascend rapidly. However, the walls of the vessel must be strong enough to withstand the external pressure and prevent it from being crushed; and mechanical grabs or delicate manipulator arms are needed for collecting samples.

The world's first ever working submarine was built around 1620 by Cornelius van Drebbel, although designs for underwater vehicles had been drawn up earlier – by Leonardo Da Vinci among others. Drebbel was far ahead of his time. Little further progress on submarine craft was made until the mid-1800s when steam-driven submarines, known as the Davids, were used in the American Civil War. Deep-sea exploration had to wait much longer. The first type of submersible built to withstand the immense pressures found at great depths was the bathysphere – a hollow steel sphere, with very thick walls, which was lowered down on cables from a surface vessel. In such a steel ball, only 1.4 metres (4.5 feet) in diameter, William Beebe and Otis Barton made a record-breaking descent to 923 metres off Bermuda on 15 August 1934. But the bathysphere could only go straight down and straight back again, providing no more than a tantalizing glimpse of the ocean depths.

The bathyscaphe, invented by the Swiss scientist Auguste Piccard in the 1940s, revolutionized underwater exploration for it was fully manoeuvrable and independent of the mother ship. The name derives from the Greek *bathys*, meaning deep, and *scaphos*, ship. It worked rather like a balloon in reverse. An upper lightweight float (filled with 60,000 gallons of petrol) allowed the vessel to ascend, while ballast was added to cause it to sink: jettisoning the ballast on the bottom allowed the bathyscaphe to sail back up to the surface. Below the float was suspended a spherical steel cabin with very thick walls, containing the crew. Jacques Piccard, the son of Auguste, together with Don Walsh, a US naval lieutenant, landed on the sea floor at the bottom of the Mariana Trench on 23 January 1960 in the bathyscaphe *Trieste*. At 10,914 metres this is the deepest place on Earth (it is over 6 miles down) and the pressure is a massive 1100 bar (16,500 pounds per square inch). No one has yet equalled their record, although a Japanese robot submersible called *Kaiko* touched bottom there in 1995.

The *Trieste*'s voyage proved that people could descend to the ocean floor and return unharmed and its success spawned a new generation of submersibles in which the cumbersome flotation tank was replaced by a pressure hull that provided the primary buoyancy. Japan, France, Russia and the United States all now possess their own submersibles. Perhaps the most famous of them is *Alvin*, launched by the Woods Hole Oceanographic Institute in 1964, which was used to locate a hydrogen bomb accidentally dropped in the Mediterranean sea off the coast of Spain, to discover the hydrothermal vents at the mid-ocean ridges, and to find the wreck of the *Titanic*. The very latest type of

The World's First Submarine

The world's first ever working submarine was built around 1620 by a Dutch alchemist, Cornelius van Drebbel (1572–1634), who was living in London at the time. He built three in total, the last being the largest and the most elaborate. Remarkably, it cruised underwater down the Thames from Westminster to Greenwich, watched by King James I. It looked rather like an oversized walnut and was covered in greased leather to keep out the water. Near-contemporary pictures suggest that the submarine was powered by oars, of which there were six on each side. What is not clear, however, is how the oars were manipulated from inside the vessel without allowing the water to seep in. Another puzzle is how the rowers (and passengers) managed to breathe. Apparently, the submarine could stay submerged for as long as an hour and a half – long enough to cause an uncomfortable fall in oxygen and a rise in carbon dioxide.

It is clear from references in contemporary accounts to the 'troubled air' that the air quality in Drebbel's machine did indeed deteriorate. How he solved the problem is less certain. One writer claims that the

Twentieth-century painting of Cornelius van Drebbel's submarine by G. H. Tweedale

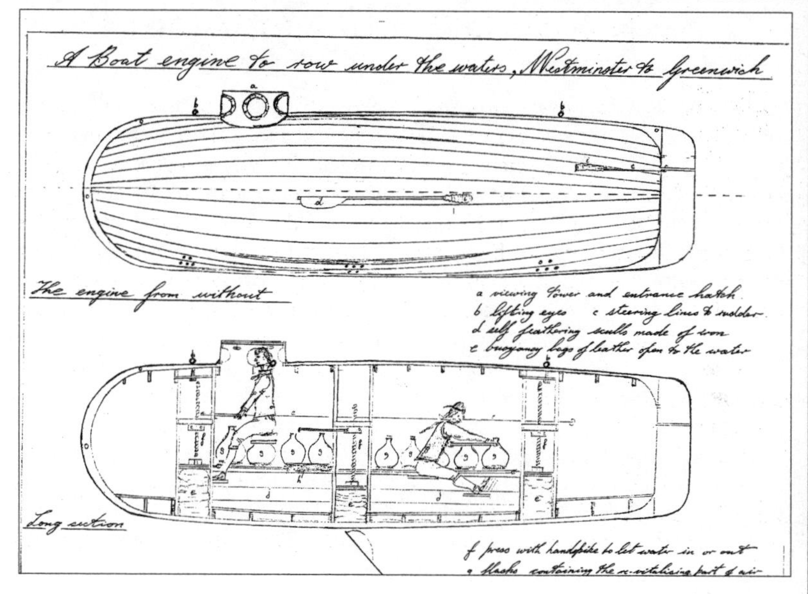

Working drawings for building a replica of Drebbel's submarine, designed by the historical boat-builder Mark Edwards on the basis of contemporary technology

submarine was connected to the surface by an air tube. However, the scientist Robert Boyle, who interviewed Drebbel's son-in-law, wrote (in 1660, some forty years later) that Drebbel 'would by unstopping a vessel full of this chymicall liquor, speedily restore to the troubled Air such a proportion of Vitall parts as would make it againe, for a good while, fit for respiration.' What the liquor was remains unclear, because oxygen was not officially isolated until 150 years later. However, one possibility is suggested by the fact that in 1610 Drebbel paid a visit to Prague where the Polish alchemist Sendivogius was working. Sendivogius had a passion for salt-petre (potassium nitrate). He described it as the 'secret food of life' and stated that the 'ariel nitre' produced when it was burnt kept people alive. His observation was quite correct, for potassium nitrate gives off oxygen when heated. So perhaps Drebbel used jars of 'ariel nitre', or even burnt salt-petre itself, to keep the air pure. The difficult question of why the carbon dioxide concentration did not rise to levels high enough to cause the rowers to become unconscious has not been solved. Perhaps, the journey was simply too short.

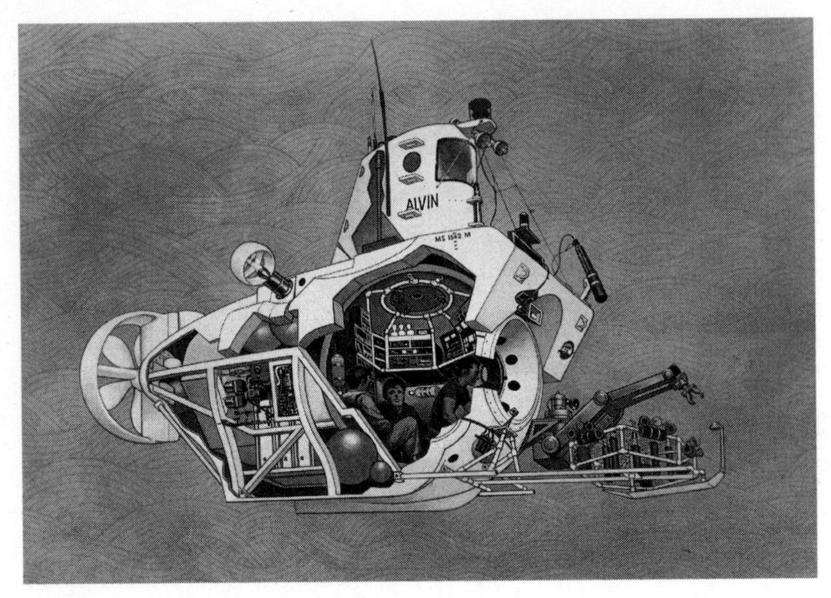

Alvin is a deep-sea submersible operated by the Woods Hole Oceanographic Institute. It has a crew of three (two scientists and a pilot). A typical dive to a depth of 4500 metres takes about eight hours, with four hours spent on the bottom.

submersible is *Deep Flight,* a fast, highly mobile craft resembling a winged torpedo, designed by Graham Hawkes, that 'flies' through the water. As yet, however, *Deep Flight* has only been tested at relatively shallow depths.

Life Under Pressure

Today, commercial divers are regularly engaged in many types of underwater work, such as pipeline surveys, oil-rig maintenance, inspecting and repairing the hulls of ships, wreck salvage and even forensic work. Many more people dive for pleasure. The depths that they can reach are dictated by the type of gas they breathe but although oxygen toxicity and nitrogen narcosis can be overcome by the use of exotic gases, eventually high pressure nervous syndrome imposes a lower limit on the free diver. In addition, divers are greatly affected by cold, and decompression sickness limits the maximum rate of ascent that is permissible. This means that divers can work safely on the continental shelves, but cannot descend below them to the abyssal plains. For deep-sea exploration, pressure-resistant submersibles or remote probes are therefore needed. There is currently considerable debate about which of these is the best option, but it seems likely that both will continue to be developed, for just as the rewards of deep-sea exploration are potentially enormous – oil and mineral wealth of immense proportions, bacterial enzymes and natural products that could revolutionize biotechnology and medicine, a unique ecosystem barely studied by scientists – so, too, is the excitement and challenge of actually seeing it with your own eyes.

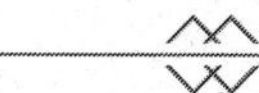

Getting into Hot Water

SOME YEARS AGO, a Japanese colleague introduced me to an oriental form of trial by fire. He took me to Ibuski, a small town in the south of Japan famous for its *onsen* (spa baths). It lies on the edge of the sea, with a magnificent view of an active volcano with the evocative name of Sakura-jima, or cherry mountain. Clad in nothing but a cotton kimono, I emerged on to a wide beach of black sand to be greeted by an extraordinary sight. Planted in regular rows across the sand, like peculiar cabbages or abandoned footballs, were hundreds of human heads. It looked as if some ancient Samurai had gone berserk and the fruits of his labours were laid out on the shore waiting for the sea to wash them away. The mystery was explained when I was beckoned by an old Japanese man clutching a shovel, who proceeded to dig me a grave. I lay down in the long shallow pit and he covered me over carefully so that only my head emerged from the sand. Being buried on the beach was not the cold clammy experience I remembered from childhood holidays in England. Water heated by the nearby volcano percolated through the shore so that the sand was hot. Its warmth enveloped me, seeping through the thin cotton garment, unteasing knots in my muscles that I hadn't even known existed. Lulled by the soft splash of the waves at my feet, I drifted into sleep. I was awakened by my Japanese friends who gesticulated to the large clock on a stick, like a giant lollipop, that dominated the beach. We had been steaming for fifteen minutes and our time was up.

We spent the next ten minutes in the adjacent buildings washing off every speck of sand, scrubbing and soaping ourselves vigorously, cleaning hair and nails and skin until everything was polished. Only now, naked and scrupulously cleansed, were we ready to enter the *onsen*, the communal hot bath.

'It's hot,' I was warned. I was unconcerned. I always take my baths hot, drink my tea boiling and am known to have asbestos fingers. I stepped boldly into the pool - and leapt straight out again. It was scalding hot. At least 45 °C. I thought I must have first-degree burns. I stared at the fine-boned Japanese women lying in the pool. How could they bear it? They smiled and nodded encouragingly, chattering to each other in high flute-like voices. I could not understand how they were not cooked. Thoughts of cannibals' cooking pots and the tortures endured by those accused of witchcraft in the Middle Ages swam through my mind. Gingerly, I inched into the water, trying to ignore the heat, and stretched out my arms along the edges of the bath to provide a large surface for evaporative cooling. I looked around. It was rather like sitting in a giant conservatory filled with tropical plants and many different pools. I was reminded of the place between the worlds in the Narnia stories, where each pool led to a different world. Here, they were filled with water at different temperatures and mineral content. When I emerged from my pool five minutes later I was a bright cherry red, like a boiled lobster. All my blood had been directed to my skin as my body tried desperately to cool down – to no avail, for not only could I not get rid of the heat I generated myself, but I was rapidly accumulating that of the bath. I sat on the edge of the pool, my skin pouring sweat. But I felt marvellous. The heat had dissolved all the aches of body and mind. Henceforth, wherever I went in Japan, I sampled the local *onsen*.

One of the most memorable of these experiences was a winter visit to an *onsen* high in the Japanese Alps. This was Mount Zao, where the poet Basho made a pilgrimage that inspired some of his most famous haiku. Snow cloaked the trees so thickly that their shape was obscured and they looked like melted candles. Shadowy grey mountains marched range after range into the distance, their shapes softened by wisps of cloud. It was the soft serene landscape of a Japanese painting – all black and white and shades of grey, with an ethereal oriental beauty that I had thought existed only in the imagination of the artist but now recognized with surprise was in fact a realistic portrayal. Small timbered houses clustered on the edge of the mountain, nestling in deep snow. Between them, hot streams ran steaming through the streets, enveloping the unwary passer-by in a warm sulphurous cloud.

The *onsen* was an ancient stone bath, partly protected by a wooden verandah, but otherwise open to the elements. It was fringed by a Japanese garden and enjoyed glorious views across the mountains. Water ran continuously through the pool from a natural hot stream. The icy air froze us as we walked naked to the *onsen* across the snow – this time I was glad of the bath's heat. I was less happy with the steam that billowed from the water. It smelt strongly of sulphur and caught at my throat. Lying in the water, half-hypnotized by the heat, I asked my companion to translate a small notice on the wall. It did not say 'No smoking', as I had assumed, but instead advised us to wash well after using the *onsen* because the water was so acid that it would rot our clothes. I wondered idly, in my soporific state, what it might do to my skin. But, in fact, it is the intense heat that is the danger, for although a short dip is marvellously invigorating, to remain there too long would, quite literally, be fatal.

3

LIFE IN THE HOT ZONE

'My strength is dried up like a potsherd;
and my tongue cleaveth to my jaws;
and thou has brought me into the dust of death.'

Psalm 22

One morning towards the end of the eighteenth century, the Secretary of the Royal Society of London, one Mr Blagden, ventured into a room heated to 105 °C, taking with him some eggs, a piece of raw steak and a dog. A quarter of an hour later, the eggs were baked hard and the steak cooked to a crisp but Blagden and his dog walked out unharmed (although the dog had had to be kept in a basket to prevent its feet from being burnt). Their ability to endure a temperature above the boiling point of water seems even more remarkable when one considers that proteins are denatured, and cells begin to be irreversibly damaged, as the temperature rises above 41 °C; that a body temperature of 43 °C is lethal for humans; and that almost all cells are killed if their temperature exceeds 50 °C for a few minutes. Yet as Mr Blagden so vividly demonstrated, the human body can survive exposure to 105 °C for almost fifteen minutes. How this is the case, is the subject of this chapter.

Our lives depend on a nuclear reactor 93 million miles away, which both lights and warms our planet. The sun has a surface temperature of 6300 °C. That of the Earth is a great deal cooler, but it can still reach levels that humans find difficult to tolerate. The hottest air temperature ever recorded on Earth is a searing 58 °C, measured in the shade in El Azizia, Libya, in September 1992. Temperatures over 45 °C are recorded routinely during the summer months in central Australia, the Gulf states and Sudan, and objects exposed directly to the sun can heat up even more, so that metal becomes too hot to touch and sand burns the feet. The effects of solar heating can also be considerable in cold environments. The sun warms the snowfields of Everest to 30°C, polar explorers may suffer simultaneously from sunburn and frostbite, and even in the freezing vacuum of space objects exposed to the sun's rays heat up fast.

The highest temperatures on Earth are found in deserts. The definition of a desert is a place with less than 10 inches of rain a year, but many deserts receive far less and in some of them it may not rain for years on end. The lack of clouds means that radiation from sun and sky is intense, causing the air and ground to heat up rapidly in the day and cool equally dramatically at night. Water is scarce, so that for most of the year the ground is barren, but the hot still air at midday generates mirages that transform the parched earth into shimmering phantom lakes. The heat can be extreme and is exacerbated by hot dry winds, which rob the body of water, shrivelling the skin and dehydrating the nasal passages. Sand and dust blown around by desert winds are abrasive and can choke. Ultraviolet radiation causes sunburn, and the harsh light produces a dazzling glare. It is not a pleasant climate for humans. Yet people have made their home in the desert for centuries and thousands of visitors come each year to witness its spectacular beauty – the great wind-carved dunes and gloriously coloured sculpted rocks. A combination of behavioural and physiological adaptations ensures their survival.

Body Heat

To understand how humans cope with extremes of heat, it is helpful to consider first what we mean by body temperature, and how it is regulated under normal conditions. Not all areas of the body are maintained at the same temperature. What is commonly meant by body temperature is more truly defined as the core temperature – that deep within the tissues of the chest and abdomen. This is kept at around 37°C, although it shows a diurnal fluctuation of about half a degree, being higher in the late afternoon and lowest just before dawn. In women, the core temperature is also affected by the menstrual cycle; it rises just before ovulation and remains high from the fifteenth to twenty-fifth days of the twenty-eight day cycle. This variation enables a woman to determine when she is most fertile and is exploited in the rhythm method of birth control.

As a thermal imaging camera clearly demonstrates, the temperature of the outer shell of the body can vary widely from that of the core. A nude person in a cold room may have a skin temperature as low as 20°C and their arms and legs will also be colder than the body core. Conversely, during heavy exercise the temperature inside the working muscles can rise to 41°C, although the core temperature may only increase by one or two degrees. Areas of high blood flow are also warmer, which is why your face feels hotter when you blush.

The normal limits of the core temperature are 36 to 38°C; clinically, hypothermia is defined as a temperature below 35°C and hyperthermia as one above 40°C. Above a core temperature of 42°C, death occurs from heatstroke. Thus, although humans can, under special circumstances, survive extreme cooling (see Chapter Four), an increase in their core temperature of as little as 5°C is fatal. Sperm appear to be particularly sensitive to high temperatures, much more so than the rest of the body, which is one reason why in mammals the testes are located outside the body, where they are kept cooler. It is ironic that although tight trousers may seem sexy, they actually reduce a man's fertility because they decrease heat loss and thereby sperm production.

Feeling the Heat

The question of how the body senses its internal temperature has occupied scientists for many years. Subjectively, it is clear to us all that the nerve endings in the skin produce the conscious sensations of warmth and cold. Yet a moment's thought suggests that the temperature that matters for our survival is not that of the skin, but rather that of the brain. It would therefore be more logical to monitor the brain's temperature rather than that of the body surface, just as our central heating systems are controlled by a central thermostat rather than by many hundreds of individual ones on the outside walls of the house.

The body's thermostat was discovered by E. Aaronsohn and J. Sachs in 1885. It is located in the hypothalamus, an area of the brain that lies at the base of the skull. Even long after its discovery there was still controversy about whether the brain or the skin is more important in temperature control. The issue was finally settled by implanting a temperature sensor within the brain of a volunteer scientist and then testing whether his body's response to a cold stimulus was determined by the temperature of the brain or that of the skin. In order to cool the blood reaching the brain rapidly, without affecting the skin, the subject was given ice cream to eat. The fact that this elicited the typical response to cold clinched the argument: the master controller of body temperature lies in the brain.

But temperature sensitivity is not confined to the brain. You have only to drink a cup of scalding hot coffee inadvertently, and then spill it over yourself as you leap up with the shock, to appreciate that your skin, tongue and the lining of the mouth and throat also possess heat sensors. These do not detect the actual temperature of our surroundings, but rather that of the skin in which they are embedded. A simple

The Development of the Thermometer

Galileo Galilei, more famous for his observations with a different instrument, the telescope, was the first to produce a thermometer, around 1610. Galileo was professor of mathematics at Padua University and to supplement his rather meagre income he manufactured and sold scientific instruments. His thermometer was simply a long hollow glass tube, partially filled with water, which was sealed at one end and immersed in a beaker of water at the other (some authorities suggest it was actually wine). When the temperature increased, the air in the tube expanded, pushing the water down the column; the higher the temperature, the lower the water level. By engraving a scale on the side of the tube, a quantitative measurement could be made. The chief difficulty with this device was that the level of liquid was also sensitive to the air pressure, so that it fluctuated even when the temperature was constant. This problem was solved by sealing the other end of the tube.

The next major advance was made by Gabriel Daniel Fahrenheit, a German scientific instrument-maker working in Amsterdam, who introduced the use of mercury rather than water (or alcohol) in thermometers, in 1724. Mercury has the advantage that it expands more evenly with temperature, does not evaporate and is easier to see. Fahrenheit's temperature scale was a modification of that used by the less well known scientist Réaumur and was based on three fixed points: the freezing point of water (32°F), the boiling point of water (212°F) and the armpit of a healthy male (98.4°F). It is still used today in the United States. Fahrenheit was also one of the first to report that the boiling point of water varied with the barometric pressure.

In addition to Fahrenheit and Réaumur, a number of other people experiment illustrates this fact – air blown over the hands by an electric hand-drier feels cool while your hands are still wet but becomes uncomfortably hot once they are dry.

The temperature sensors in our skin come in two different varieties. One type responds to temperatures between 13 and 35 °C and signals the degree of cold or warmth. These are known as cold receptors because the rate at which they send electrical signals to the brain increases when the temperature falls. They are most sensitive around 28 °C, which suggests that humans may have evolved in an environment with about this mean temperature.

The other kind of receptor is stimulated by heat, and is felt as pain.

invented thermometers, but different scales were used and it was widely believed that the same fixed points would not apply at different places in the world. Anders Celsius clarified the muddle in 1742 when he invented the 100-degree thermometer scale. He worked at Uppsala University, the oldest in Sweden, and his thermometer can be found in the museum there. It still bears his own handwritten scale. Using this instrument, Celsius showed that snow always melted at the same point on the scale, whether it was in the northern wastes of Lapland or the more clement climate of southern Sweden. Furthermore, using one of Réaumur's thermometers, he demonstrated that the freezing point of ice in Sweden was the same as that measured in Paris by Réaumur. Celsius fixed 100 °C as the melting point of ice and 0 °C as the boiling point of water, but after his death the scale was reversed to give that we know today.

Many years after these early pioneers, the British physicist Lord Kelvin (1824–1907) invented the temperature scale used by scientists. This scale starts at absolute zero, the coldest of all temperatures. Absolute zero is defined as 0 °K (Kelvin), and corresponds to –273 °C.

The first person to measure the body's temperature in a scientific way was the Venetian Santorio Santorio, who published a major medical textbook, the *Ars de Medicina Statica,* in 1612. He adapted Galileo's instrument to measure changes in the temperature, not of the air, but of the body. His instructions state: 'The patient grasps the bulb, or breathes upon it into a hood, or takes the bulb into his mouth, so that we can tell if the patient be better or worse.' Santorio also added a scale, but this was to enable the physician to determine if the patient's temperature deviated from the value measured when he was in good health, rather than to compare it to a fixed 'normal' value. In Santorio's day, it had not yet been recognized that all humans have a similar normal temperature.

These receptors have recently been isolated, and their DNA sequence determined, by exploiting their high affinity for the spice capsaicin, the active ingredient of chilli peppers. Tucked innocuously away inside brilliant red fruits, capsaicin explodes within the mouth like a volcano, creating the powerful burning sensation only too familiar to anyone who has eaten Indian or Mexican food. Attempts to douse the fire with water only succeed in spreading it further around the mouth. The initial pain is often followed by an outbreak of sweating, as if the spice had indeed elevated the body temperature.

Capsaicin interacts with the same membrane protein that is involved in the sensation of painful heat, which may explain why it is perceived as hot. The capsaicin receptor is also activated by resiniferatoxin, a toxin from the spurge *Euphorbia resinifera* that is responsible for the powerful burning sensation and skin irritation induced by the milky sap of these plants. People who regularly consume very spicy food become desensitized to the effects of capsaicin and such stalwarts are able to consume red-hot curries without apparent distress. It is possible that prolonged exposure to capsaicin leads to a reduction in the number of its receptors. Another, rather more alarming, hypothesis is that the pain-sensitive neurons may actually be destroyed, since high concentrations of the drug cause the death of cultured nerve cells in the laboratory. Whatever the reason, the desensitization of pain nerve fibres by capsaicin is the basis for the use of the spice as an analgesic for arthritis (it is applied topically, as a cream).

The capsaicin content differs between different varieties of pepper. This fact stimulated Wilbur Scoville, in 1912, to devise a way of calibrating the potency of the spice in order to provide a means of standardizing the quality of imports to the United States. His test involved measuring how much an extract of the pepper must be diluted until it was only barely detectable when placed on the tongue. On the Scoville scale, the mild bell pepper has less than 1 heat unit, the hotter jalapeno has 1000, the fiery habanero has 100,000 and pure capsaicin has a massive 10 million heat units.

Just as chilli stimulates hot receptors, so other chemicals interact with the receptors that sense cold, fooling the body into thinking the substance is cool. Menthol, the chief constituent of peppermint oil, is one example. Menthol was once believed to have considerable medicinal properties and in the 1930s more than 500 acres of peppermint (*Mentha pipertita*) were grown around Mitcham in England. Similar plantations were to be found in France, the Piedmont region of Italy, and elsewhere in Europe. The Japanese were also convinced of its value and used to carry menthol around with them in little silver boxes that hung from their belts. It is still used today in cigarettes to provide a 'cool' smoke, and in chewing gum and toothpaste to provide a 'fresh' taste.

Signals from the hot and cold temperature sensors in the skin can produce local effects. If you plunge your hand into cold water it will flush red as more blood is diverted there to warm the skin, although your core temperature has not altered. More importantly, the signals are also transmitted to the brain where the information is integrated with that from the central thermoreceptors in the hypothalamus and used to regulate the overall rate of heat production and heat loss by the body.

Unlike humans, a number of animals have specialized heat-sensing organs that are able to detect infra-red radiation and function as natural thermal-imaging cameras. The best studied are those of snakes. Pit vipers, such as rattlesnakes, have two heat-sensitive 'eyes' known as pit organs located one on either side of the head. They consist of a tiny pinhole entrance, that widens into a larger cavity several millimetres in diameter. The pit organs are able to detect the location of warm-blooded prey and enable the snake to strike accurately even in the dark. It is still not clear exactly how the pit organs work, due in part perhaps to the fact that pit vipers tend to be aggressive and their bite is deadly. Boa constrictors, anacondas and pythons also have heat sensors of extreme sensitivity – that of a boa constrictor is capable of detecting, almost instantaneously, as little as one ten-millionth of a calorie per square centimetre. This is equivalent to detecting the heat given out by a 100 Watt lightbulb (or, in fact, a person) at a distance of about 40 metres. Specialized infra-red sensors are also found on the underside of the fire beetle *Melanophila* which lays its eggs in freshly burnt wood. The adult beetles are attracted to forest fires in large numbers, being guided to their destination by the heat. They are so sensitive that they can detect a fire up to 50 kilometres away.

Firewalking and All That

Fire is a wonderful friend and a deadly foe. A child quickly learns that its brilliant leaping flames signal danger. Fear of the 'burning fiery furnace' has been used by numerous religions to ensure a supplicant's

Creatures of Fire

The phoenix was a fabulous Arabian bird, named for its gorgeous reddish-purple colour, that reputedly lived for more than 500 years. When close to death, it built a funeral pyre scented with frankincense and myrrh, faced towards the sun and burst into flames. Nine days later, a new phoenix arose from the ashes of the old. In ancient times, the phoenix provided powerful support for the idea of the Resurrection of Christ, for if a mere bird had the ability to die and rise again, how could one doubt that the Son of God could do so?

The origin of the phoenix myth is less clear. T.H. White suggests that it may have arisen from the ceremonial sacrifice of a purple heron by the Egyptian priests of Heliopolis because the sacred symbol for the sun, which died at night and was born once again the next morning, resembled a heron. Another idea, however, is that is may have arisen from the fact that some members of the crow family occasionally crouch at the edge of a small fire and spread their wing feathers into the cooler part of the flames. It is thought that this behaviour may serve to burn off parasites, while the skin of the bird is protected from the heat by its feathered coat.

The phoenix is merely a myth, albeit a glorious one. The fire salamander, however, is a real creature, with a moist glistening skin dramatically blotched in bright yellow and black. This magnificent amphibian was regarded with a mixture of horror and awe in ancient times for it was thought to be highly toxic and to extinguish fire. Because it only ever appeared in daylight after a heavy storm, it became associated with wetness and damp and this, together with the fact that salamanders were seen emerging from damp logs placed on the fire, may have led to the ancient belief that it quenched fire. *The Book of Beasts*, a Latin bestiary of the twelfth century, states:

'the salamandra has its name because it prevails against fire . . . This animal is the only one which puts the flames out, firefighting. Indeed, it lives in the middle of the blaze without being hurt and without being burnt – and not only because the fire does not consume it, but because it actually puts out the fire itself.'

Aristotle made a similar contention. Pliny was more of an experimentalist, for he tested the hypothesis by putting a salamander into the flames. Of course, the unfortunate creature was burnt to a cinder; but, despite the evidence of his eyes, Pliny continued to promulgate the myth that the fire salamander was able to quench fire.

In a footnote to his wonderful

translation of *The Book of Beasts*, T.H. White tells us that the 'Emperor of India' had a suit made from a thousand salamander skins, that Pope Alexander III had a tunic made of them and Prester John a robe. Presumably they believed, as Caxton did, that 'This salamander berith wulle, of which is made cloth and gyrdles that may not brenne in the fyre.' Indeed, when asbestos was discovered, it was assumed to be the wool of the fire salamander.

The bombardier beetle is distinguished not for its tolerance of fire, but for its use of heat as a defensive weapon. When startled, it squirts a highly caustic spray of superheated steam, laced with hydrogen peroxide, at its unsuspecting attacker. The noxious vapour is produced by a pair of glands in the beetle's abdomen, each of which has two compartments. One of the chambers is filled with an aqueous solution of the chemicals hydrogen peroxide and hydroquinone; the other contains a mixture of enzymes. When the beetle is alarmed, it injects enzymes from one chamber into the other; these catalyse a heat-emitting reaction between hydrogen peroxide and hydroquinone, and the energy generated by the reaction heats the solution to boiling point. By swivelling the tip of its abdomen, the beetle is able to spray its assailant with pinpoint accuracy. The vivid black and orange colours of the insect, and the audible explosion that accompanies each irritating discharge, help to remind its enemies that this is a beetle to be avoided.

submission, both in this world and the next. The Spanish Inquisition believed that death at the stake was necessary to cleanse unrepentant sinners of their sins, and thus save their souls from damnation, while the mere mention of Hell conjures up images of everlasting fire. Our fascination with the ability of fire-walkers to walk barefoot across a bed of hot coals without harm derives not only from the imagined pain but also from these cultural associations. Indeed, fire-walking may have begun as a means of assessing the guilt of the sinner, or testing the sincerity and spiritual strength of the novitiate.

There is nothing supernatural about fire-walking, however, nor does it require a special 'state of mind'. The secret lies in the low thermal conductivity of wood and the relatively short time that the fire-walker's feet remain in contact with the hot coals. Wood is a very poor conductor of heat (which is why saucepans used to have wooden handles) and charcoal is almost four times better as an insulator. This means that little of the heat in the hot ashes is passed on to the feet and it is possible to walk across embers as hot as 800°C for as far as 52 metres. Fire-walking is thus a matter of physics rather than physiology.

With protective clothing, humans can tolerate extreme heat. The military have 'fear naught' suits made of several layers of felted wool. Originally developed to shield stokers from hot sparks, they were later adapted to protect soldiers from flash burns or brief blasts of intense heat caused by explosions. Wearing 'fear naught' gloves, a man can even pick up a bar of red-hot metal. Flame-resistant synthetic materials, like Nomex, are used by racing drivers and oil-rig personnel to prevent burns in flash fires, and stunt-men wear suits of similar material to film scenes in which the actor is consumed by a sheet of flame. Protected by such a suit, it is possible to remain cool in a conflagration for several seconds.

Without protection, even moderate heat kills cells. Accidentally touch a hot iron with your finger and your flesh is singed white because the skin cells are killed. Such minor surface burns cause death of the upper layer of skin cells. If exposure to heat is prolonged, injury to underlying tissues may also occur. Damage can continue even after the affected area has been removed from the fire because of heat stored in the tissues, which is why rapid cooling with cold water or an icepack is the best treatment for minor burns.

Although all mammalian cells die when heated above 50°C for a few minutes, humans can tolerate much higher air temperatures for brief periods provided that the air is very dry, as Mr Blagden so vividly demonstrated. Many people will know this from experience – a sauna is usually heated to around 90 °C. Experiments have shown that dry air temperatures as high as 127 °C can be sustained for twenty minutes and there are anecdotal reports of even hotter temperatures being tolerated for shorter times. This is because sweating cools the surface of the skin to a temperature considerably lower than that of the air, which explains why very hot air can singe your hair and eyebrows, yet your skin remains undamaged. Extremely high temperatures, such as those encountered in flash fires, are exceedingly dangerous, for the hot air damages the delicate lining of the lungs and overcomes the skin's cooling system, resulting in severe burns. Fortunately, air temperatures on Earth seldom exceed 50 °C and heat so intense that it burns the skin is normally only encountered during a fire.

Despite the fact that people can survive dry air temperatures higher than boiling point for short intervals, this cannot be tolerated for ever. With time, the body temperature inevitably rises. Brain cells are extremely sensitive to heat – 42 °C is all they can bear – and an increase of a few degrees in the temperature of the blood can have a profound effect on brain function. In the long run, therefore, our ability to cope with heat relies on thermoregulatory systems that ensure the body temperature is kept below 42 °C.

Hot Stuff

Heat is a by-product of life, as is evident from the rapid cooling of the body after death. As the philosopher John Locke wrote in about 1666, 'no one when he stops breathing gets hotter'. The biochemical reactions that power our cells are not completely efficient and, like a car engine, they generate a small amount of heat as a by-product. At rest in a warm climate, the heat produced by the body is sufficient to supply the interior warmth we need, but in a cold climate the heat loss to the environment can be so great that supplementary warmth is required. By contrast, physical exercise can stimulate heat production more than five-fold and a substantial increase in heat loss is essential. There are also many places in the world where the ambient temperature is greater than that of the body and heat gain from the environment must be minimized.

Before there was a way to measure body temperature, it was thought that it varied in different parts of the world, and that people who lived in the tropics had warmer body temperatures than those who lived in the frozen north. Johannis Hasler, writing in 1578, even provided a chart of how much heat or cold was expected in people living at different latitudes. In medieval Europe, medical practice was based on the classical theory of Galen which postulated that the body contained four cardinal humours (the name derives from the Latin *umor*, meaning fluid). These consisted of blood, phlegm, black bile (melancholer) and yellow bile (choler). Each person's temperature, a word that was used synonymously with temperament, was determined by an individual mixture of these humours. A predominance of blood produced a sanguine temperament, of phlegm a phlegmatic nature, of black bile a melancholic person and of yellow bile a choleric individual. Providing their humours were in balance, an individual was healthy. Because each person's balance was unique, it followed that there was no normal temperature for the body and thus what might appear to be a fever in one person could be quite normal for another. 'It is evident,' Sir Walter Raleigh observed in 1618, 'that men differ very much in the temperature of their bodies.' Similarly, Sir Francis Bacon commented that 'there are persons of all temperatures'. The fact that we continue to speak of hot-blooded and cold-blooded individuals, and of a person's temperament, is a legacy of this ancient belief.

Like other mammals, however, humans are homeotherms and maintain a stable body temperature regardless of the external temperature. This means that the rate of heat production must be balanced with that of heat loss. Life in the hot zone is thus a matter of reducing heat production and increasing heat loss. The third way – storing heat within the body for later disposal, by allowing your core temperature to rise – is not an option for humans, but is sometimes employed by other animals, as discussed later.

Chilling Out

All animals, including humans, reduce heat stress by behavioural adaptations, which include inactivity and seeking the shade. Food intake is reduced, because metabolism generates heat, and food with a high water content becomes increasingly attractive. Ices, fruits, cucumbers and long cool glasses of lemonade are preferred summer fare. Because muscle activity generates considerable amounts of heat, manual work is confined to the beginning and end of the day, when it is cooler. Many people take a prolonged midday siesta – except, of course, for those famous mad dogs and Englishmen, lampooned for their love of the midday sun. But Nöel Coward's song is not mere fiction. The British Raj in India actually believed that physical exercise was essential to prevent them succumbing to tropical diseases and everyone, men and women alike, took some sort of exercise in the afternoon. They would have appreciated the current credo of physical fitness. Galloping around in the midday sun playing polo or tennis, however, put them at risk from heatstroke.

Humans also regulate their clothing, housing and degree of exposure in hot climates. Unlike tourists, indigenous peoples of the desert often wear several layers of loose-fitting clothing that completely cover the body. Likewise, camels and other desert animals have thick fur, particularly on their backs. Intuitively, this may seem confusing at first but there is a simple explanation. Fur and clothing are very effective heat shields, providing an insulating layer whose function is to keep the heat out when the surroundings are hotter than the body. A shorn camel requires far more water because it gains heat more rapidly. And far from providing relief, taking off your clothes causes you to heat up faster. Loose clothing is best, for it enables circulating air currents to evaporate sweat while at the same time providing protection from the hot desert sun.

Animals have evolved many remarkable behaviours to avoid heat stress. The Namibian toad, one of the few desert amphibians, sits out the day buried beneath a few inches of sand where the temperature is much lower than that on the surface, and only emerges in the cool of the night. Honeybees make use of a different strategy. They employ evaporative cooling to maintain the temperature of their developing larvae at a constant 35 °C. If it gets too hot inside the hive, they spread droplets of water over the surface of the comb and then fan their wings to create air currents that replace the hot moist air with cooler dry air. Other animals survive the intense summer heat by adopting a state of extreme torpor known as aestivation, in which the metabolic rate is drastically decreased. Tucked away in a shady spot or a cool underground burrow, they simply wait it out until times get better.

In the days before air-conditioning became widespread, humans also built subterranean dwellings to escape the heat: the Mughals retreated to cool *tykhana* (basements); the Matmata houses of the Sahara are sited 10 metres below ground; and people in the Australian desert town of Coober Pedy, famed for its opal mines, also lived in underground homes (some still do). Even in less torrid climates, vernacular architecture reflects the need to reduce heat stress. Windcatchers once graced the roofs of Hyderabad in Pakistan, scooping the cool afternoon breeze into the room. Traditional Japanese houses are designed so that the walls can be slid back to expose the house to cooling through-winds. And in the Dorsetshire countryside where I grew up, the cottages were made with walls of baked clay and straw, known as cob, that were often up to two feet thick. In the hot summer days of my remembered childhood their insulating qualities provided a cool refuge from the heat.

Like most desert people, the Tuareg wear long robes that cover them completely.

Sweating It Out

While your behaviour can reduce the rate at which you soak up heat from your surroundings, heat produced by the body itself must be removed. The skin serves as the main organ of thermoregulation in humans. Heat generated by the muscles and other internal organs is carried by the blood to the skin, where heat loss to the environment is regulated by varying the amount of blood flowing through a network of fine blood vessels that runs close to the body surface. A rise in body temperature produces dilation of these superficial blood vessels and shunts warm blood closer to the surface of the skin, increasing heat loss. This accounts for the flushing of the skin that occurs when you are hot. Conversely, when the body temperature falls, the surface blood vessels constrict and blood is preferentially channelled through deeper vessels so that heat is conserved. The thermoregulatory system of the body is simply a more sophisticated version of the engine-cooling system of a car, in which the heart replaces the water pump, the blood serves as the circulating coolant and the skin functions as the radiator.

Heat is lost from the skin by four processes: radiation, conduction, convection and the evaporation of sweat. At rest in still air, radiation accounts for around 60 per cent of heat loss, with convection and conduction contributing about 20 per cent (more when there is a wind). As long as the skin temperature is less than that of the body core, radiation, convection and conduction are sufficient to cool the body. These processes enable the core temperature to be maintained in still air of less than 32 °C.

But there are many places on Earth where the prevailing temperature is much higher than that of the body and consequently heat will be taken up by radiation and conduction, increasing heat stress. During the first Gulf War, many ships sailed to the Persian Gulf through the Strait of Hormuz. The outside air temperature was a blistering 47 °C and the humidity was very high. Under clear skies, in blinding sunshine, with light reflecting off the water, the heat was intolerable. When attired in their antiflash gear and action coveralls, the gun crews were limited to a brief ten minutes on the upper deck. Nor are civilians unaffected. Every year, thousands of pilgrims make their way to Mecca, where the average air temperature is over 40 °C. Many succumb to the heat.

The Physics of Heat Transfer

Heat is the energy of molecular motion. The temperature of a gas is determined by the average speed of its constituent molecules: the faster they whizz around, the hotter it is, and the slower they move, the colder it is. In solids, the constituent molecules are bound to each other and scientists often imagine them as collections of springs connected together: the higher the temperature, the greater the amplitude at which the springs oscillate and the lower the temperature, the smaller the amplitude of the oscillations. At absolute zero (-273°C), there is almost no oscillation at all. You might wonder why the qualifier – surely, by definition, no movement should occur at absolute zero. The reason this is not the case results from the vagaries of quantum physics which states that it is not possible to predict the position of a particle at precisely the same time as its momentum (this is Heisenberg's famous uncertainty principle). The more precisely you try to predict where a particle actually is, the more uncertain you must be about how much momentum it has, and vice versa. Consequently, Heisenberg's principle implies that the molecules of a solid must always be vibrating just a little, even at absolute zero.

Heat is transferred from one object to another by conduction, convection and radiation. Conduction describes the process in which heat is transferred between two objects that are in direct contact, such as skin and air. If these are at a different temperature, heat will flow from the warmer object to the colder one. Simply put, the molecules in the hotter object bump into those in the colder one and increase their rate of movement, while at the same time reducing their own speed. The ease with which heat flows through an object is known as its thermal conductivity: wood has a lower thermal conductivity than copper which is why copper saucepans used to have wooden handles. Insulation – the resistance to heat flow – is the reverse of conductivity. Air and feathers have a low thermal conductivity (or high insulation value) which explains why feathers with layers of air trapped between them make excellent quilts.

The transfer of heat in fluids (water or air) is increased by the process of convection. This is best explained by imagining that you are suddenly immersed in a bath of cold water. The water in contact with your skin will gradually warm up. If this water is then replaced by fresh cold water, the same process will begin again, warming further water (and further cooling your body). This process in which the water next to your skin is constantly replaced is known as convection and it results from the fact that warm water rises (because it is lighter). The temperature differences in the bath water mean there is a constant water circulation with warm water rising and

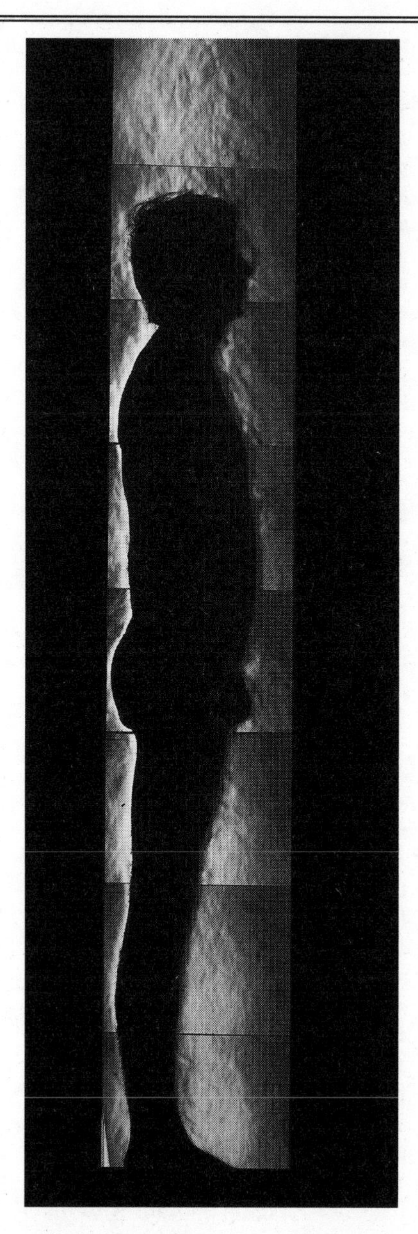

Schlieren photograph of a naked man, showing the column of rising warm air that constantly envelops us.

cold water descending, and these circulating currents constantly replace the water next to your skin and facilitate heat transfer.

Conduction and convection are simply explained, but the nature of radiation has puzzled scientists for centuries. All objects emit electromagnetic radiation and the hotter they are, the larger the amount of radiation emitted. This radiation is emitted over the whole range of the electromagnetic spectrum, but the peak of the emission depends on the surface temperature of the object, and shifts towards shorter wavelengths when the object heats up.

The wavelength determines whether or not we can see the radiation as colour or feel it as heat. Long wavelength radiation is invisible and we perceive it solely as heat: for example, you can still feel the heat from a fire that has long since stopped glowing. This is known as infra-red radiation. As the temperature of an object increases, the wavelength where most radiation is emitted shifts into the visible range and the object starts to glow. First manifest as a dull red, the colour shifts from red through yellow to white (hence the phrase 'white-hot') as the object gets hotter and the average wavelength shortens. You might expect that the colour change should follow the colours of the spectrum and shift from yellow through green to blue. Yet it is clear simply by heating an iron poker in the fire that this is not what happens. This is because (as noted above) the poker is simultaneously emitting light over the whole range of the electromagnetic spectrum and it is only the wavelength at which the *peak* emission occurs that shifts with temperature. Furthermore, the total amount of radiation emitted increases dramatically with temperature so that much more long-wavelength radiation is also emitted. Thus, the light emitted by the poker is a mixture of different wavelengths and hence appears white, like sunlight, and a white-hot poker feels a great deal hotter than a dull red one or the dying embers of a fire.

The sun has a surface temperature of about 6300°C and emits visible radiation at a peak wavelength of around 0.5 micrometres, which is why it appears blindingly bright. It also radiates at longer wavelengths and supplies the heat that supports all life on Earth. A human body, which has a temperature of 37°C, radiates at a peak wavelength of 10 micrometres, which is well outside the range of visible light. However, if the environment is well insulated, it is possible to feel the heat radiated by another human being (for example, in bed). Incidentally, it is worth noting that the temperature of the sun is about twenty times that of the human body on the Kelvin scale (6600°K as against 300°K) and that the peak wavelength of the radiation emitted is around twenty times shorter than that of a human, which demonstrates that the peak wavelength is simply proportional to temperature.

Like light, heat can be thought of both as a wave or as particles (known as photons). To understand how radiant heat transfer occurs – and why it can cross the vacuum of space from the sun to the Earth – it may be helpful to consider heat as photons absorbed or emitted by the atoms of your body. An atom is like a miniature solar system. At its heart is the nucleus and orbiting around it are one or more electrons. The electrons are spaced at discrete intervals from the nucleus, rather like the planetary orbits. At this point the analogy falls apart, for the orbital in which an electron finds itself depends on its energy and electrons may leap between the different orbitals depending on whether they absorb or emit energy. We may think of this energy as photons, or light particles. Jumping up to an outer orbital is evoked by the absorption of a photon, while a photon is emitted when the electron falls back down to a lower orbital.

Molecules absorb and emit radiation in different ways to atoms: they increase or decrease their amount of vibration. Photons travel though a vacuum at the speed of light – 186,000 miles per second. Those arriving from the sun are absorbed by the molecules in our skin, enhancing their vibration and warming us up. Heat is lost by radiating photons when the amount of molecular vibration decreases. As you read this, your body is radiating photons to the world around you. You are in a constant silent dialogue, exchanging photons with the people and objects in the room in which you sit.

When the air temperature is greater than that of the body, the only way to lose heat is by sweating. This works on the same principle as a terracotta wine cooler and exploits the fact that the conversion of liquid water into water vapour requires a great deal of heat. At body temperature, about 2400 calories are used to evaporate each millilitre of water – roughly the same as needed to heat the same amount of water from freezing to boiling point.[1] Most of this heat comes from the body itself, so that the evaporation of sweat cools the skin. Consequently, blood flowing through the skin is chilled, and on circulation to the warmer core, helps reduce the body temperature.

We have about 3 million sweat glands, about half of which are found on the skin of the chest and back. Large numbers also occur on the forehead and palms of the hands. It is actually quite easy to see the individual pores by coating your skin with suntan oil and sitting in the hot sun for a few minutes. As the skin heats up, tiny beads of sweat appear, each one at the mouth of a single sweat gland. The oily film reduces the rate of water evaporation and makes the sweat easier to see (a hand lens makes it even clearer).

Sweating is stimulated by the hormone adrenaline, which is released when the body temperature rises. Adrenaline release is also increased by stress, which explains why we suffer from sweaty palms and moist foreheads when afraid. There is an old adage that states 'horses sweat, men perspire and ladies gently glow'. Although generally considered to be a Victorian nicety, there is in fact some truth in this saying as women produce only about half as much sweat as men when exposed to the same amount of heat. There is also considerable variability between different races: natives of New Guinea, for example, sweat less than Nigerians or Swedes.

Sweating can increase heat loss almost twenty-fold but only at the expense of substantial water loss – as much as 3 litres per hour. However, such high sweat rates cannot be sustained for long periods and the usual rate of water loss for someone working in the heat is around 10–12 litres a day. In dry desert air, sweat may evaporate so rapidly that the skin appears dry – place the palm of your hand on your arm, however, and you will find that it is quickly coated with sweat. Even if you do not feel hot, evaporative heat loss will be occurring, at a rate of around 0.8 litres of water each day.

Evaporative cooling is of considerable importance to athletes. Cyclists in the gruelling Tour de France are able to cycle continuously uphill for periods as long as twelve hours at a stretch. In the laboratory, they are often surprised and chagrined to find that they are unable to maintain the same rate of exercise for even an hour. On the road, the head wind caused by their forward movement rapidly removes the air layer next to their skin and markedly enhances evaporative cooling, but on a static exercise bike this convection is greatly reduced and the rate of heat loss is correspondingly less, so the cyclist quickly becomes exhausted. If an artificial breeze is created by turning on a fan, however, they are able to continue for far longer. The sudden reduction in evaporative cooling may underlie a number of incidents in which a cyclist or runner suddenly developed heatstroke after stopping exercise. It is possible that the abrupt cessation of air moving past the body decreased the rate of heat loss sufficiently to elicit a significant rise in body temperature. This is perhaps one reason for the horseman's maxim that horses must 'cool down' gradually after exercise and not stop instantly.

On a hot day, a quick dip in the pool or a brief shower leaves your skin covered in water droplets and helps you cool down by enhancing evaporative losses. Elephants employ a similar strategy, hosing themselves and each other down with water. A number of Australian animals have evolved a more labour-intensive form of evaporative heat loss. Instead of sweating, they lick themselves copiously, relying on the evaporation of their saliva to cool them down. As may be imagined, this method is not a very effective means of cooling and seems to be more of a last resort. A different approach is taken by the wood stork, which urinates on its legs every minute, thereby enhancing evaporative cooling. Closer to home, dogs hang their moist tongues out of their mouths to accelerate heat loss and pant to cause cooling of the nasal passages and facilitate evaporative heat loss from the upper airways.

Humans can live comfortably in environmental temperatures well above that of the human body providing the air is sufficiently dry. If the humidity is greater than 75 per cent, however, the sweat drips off the skin as liquid, without evaporating; in these circumstances, sweating only causes dehydration and its cooling effect is lost. This explains why the combination of high humidity and extreme heat is so oppressive. Governor Ellis, writing of the climate in the West Indies and Jamaica, reflected 'One can scarcely call it living, merely to breathe and trail around a vigourless body; yet such is generally our condition from the middle of June to the middle of September.' The Australian poet Les Murray put it more eloquently,

> '. . . we were clutched back into the rancid
> saline midnights of orifice weather,
> to damp grittiness and wiping off the air . . .
>
> Skins, touching, soak each other. Skin touching
> any surface wets that and itself
> in a kind of mutual digestion.
> Throbbing heads grow lianas of nonsense.'[2]

Most people find it difficult to bear a temperature of 50 °C when the air is saturated with moisture, yet find dry heat at 90 °C quite comfortable for short periods. Although it may feel just as hot, the temperature in a steam bath is always lower than that of a sauna. It is obvious from all this that sweating can have no effect on heat loss if you are immersed in water. This means it can be – literally – fatal to remain too long in a deep bath that is hotter than your body temperature. The hottest Japanese onsen are 46-74 °C and even the most stalwart cannot stay in for more than 3 minuites. Most people can only manage a temperature of 43 °C.

Although people usually feel exhausted on first arriving in a tropical environment, some degree of acclimatization does occur. When soldiers were airlifted from Northern Europe to Saudi Arabia during the Gulf War, they felt floppy and fatigued for the first few days. Exercise exacerbated their condition, and they quickly became exhausted – not an ideal state for an army. Within a week or so, however, the soldiers acclimatized to the heat and recovered their stamina. Acclimatization is principally due to a marked increase in the amount of sweat produced, coupled with a reduction in the amount of salt it contains.

Keeping a Cool Head

Antelope face a particular problem. They live on the hot arid plains of Africa where there is little shade, and their only means of escaping from predators is by running faster than their pursuer. But running generates large amounts of heat – up to forty times as much as when at rest. Consequently, a racing antelope is at risk from heat exhaustion.

The mammalian brain is particularly sensitive to heat, as discussed earlier, and it is the organ that dies first when the core temperature rises. One way of coping with excessive heat would therefore be to keep the brain cool but to allow the temperature of the rest of the body to rise. This is the strategy adopted by the oryx and gazelle, which tolerate body temperatures as high as 45 °C with equanimity. These animals possess a specialized vascular heat exchanger known as the *rete mirabile* (literally, the marvellous net) that cools the blood supplying the brain. Before it ever reaches the brain, the carotid artery splits up into a network of hundreds of small branches. These intermingle with a similar network of fine veins that are carrying cool blood back to the heart from the nasal passages. Heat drains from the warm arteries into the cool veins so that the temperature of the blood entering the brain is reduced, and although the body temperature may rise by over 4 °C, that of the brain changes by less than a degree. In this way, a sprinting gazelle keeps its brain cool, and stores the excess heat within its body until the crisis is over. The stored heat is dissipated at night by conduction or convection. Consequently, this strategy also conserves water because it reduces the need for sweating.

The Importance of Size and Shape

Body size is important for thermal regulation. Just as a large block of ice melts more slowly when left intact than when it is broken up, because it has a lower surface-area-to-volume ratio, so a large animal loses heat less fast than a smaller one. Tiny animals, like shrews and hummingbirds, may lose heat so rapidly they cannot maintain their body temperature at night. Conversely, large animals are in danger of overheating if they exercise in a hot climate, so that chases across the African plains are invariably brief sprints.

Ethnologists and archaeologists have long noted that the dimensions of the human body are correlated with the ambient temperature in which the different races of mankind evolved. Natural selection has moulded our bodies, so that people adapted to live in cold climates, like the Inuit of the Arctic, are short and stocky, with short arms, legs, fingers and toes. This helps to conserve heat because it ensures a low surface-area-to-volume ratio. Races that evolved in hot dry environments, like the plains of equatorial Africa, are tall and slender with much longer limbs. Not only do present-day Masai and Samburu people have this build, so too did some of the early hominids who lived in the same area of East Africa. As Alan Walker and Pat Shipman have so vividly described, the Nariokotome boy – the most complete skeleton of *Homo erectus* ever found – had even longer limbs than living Africans. Being tall facilitates heat loss as it provides a proportionately greater surface area for sweat production, while heat conduction from deep body tissues is enhanced if there is little subcutaneous fat. A long lean shape is therefore optimal for a hot climate and is of particular importance if you are a hunter and need to run to catch your food. Animals have also evolved ways of increasing their surface area to enhance heat loss. This is the chief function of the huge ears of the elephant, and the long thin featherless legs of birds.

Food is often scarce in the desert, so humans and other animals lay down fat reserves when food is plentiful. But fat is a very efficient insulator and would impede heat loss if dispersed subcutaneously. Consequently, desert dwellers tend to store their fat in a single place. The camel's hump serves this purpose – it is not, as is sometimes believed, for water storage. Likewise, the Hottentots of South Africa store fat principally in their buttocks, a condition known as steatopygia, and have long slim limbs that facilitate heat loss. Steatopygia is also common in overweight Europeans and North Americans, but is no longer of adaptive value in a well-nourished population and a colder climate.

Heatstroke

In the United States, around 250 people each year die of heatstroke and in bad years numbers may climb to over 1500. In July 1998, temperatures in the Midwest soared to over 38°C and remained there, even at night, for twenty-four days in a row. One hundred and fifty people died. During a similar, but shorter, heatwave the following year, fifty people died on one night in Chicago alone. In such extreme conditions, a person can go to bed apparently well and be found dead, or seriously ill, the following morning. Closing the windows for fear of burglars may instead precipitate a crisis from heatstroke. Old people are disproportionately at risk because they sweat less, and in the 1998 heatwave, the elderly and poor were advised to take refuge in air-conditioned shopping malls during the day. Children were confined inside and night shifts were introduced for outdoor workers.

In the early part of the twentieth century, heatstroke was considered a form of solar apoplexy. Sunlight was thought to contain dangerous 'actinic' rays that could penetrate the skull and reach the brain where they caused '*coup de soleil*'. This led to a vogue for sun helmets and spinal pads to retard the entry of solar rays. Some even advocated that a thin plate of light metal be attached to the top of the sun hat. Elspeth Huxley in *The Mottled Lizzard*, an evocative account of her life in Kenya as a young woman after the First World War, wrote that travellers wore

> 'spine-pads made of quilted cloth interwoven with a red material, and buttoned to the outside of the shirt. The sun was still regarded as a kind of dangerous wild animal that would strike you down if you did not watch it every minute of the day between nine and four o'clock.'

Her description of the arrival of Cousin Hilary presents an even more remarkable picture. He was swathed in layers of protective clothing including,

> 'an enormous topee to which was attached a long purple scarf, floating down over the back. Under this was a spreading spine pad made of kongoni hide lined with red flannel. The face was concealed by large black goggles, and over all this was a huge striped sun umbrella. Clutching the umbrella, he hurried into the shade of the veranda and cautiously began to divest himself of some of his casings. "Thatch *over* corrugated iron," he remarked. "That is a step in the right direction, but there should be two thicknesses of bituminized felting between the tin and the thatch. I think, however, that I can venture into my lighter headgear.'

Nor were his concerns confined to himself. He admonished his cousin, Elspeth's mother:

> 'Do you think it is safe to stand on this veranda without a hat? And that blouse – charming, becoming, but nothing there to keep off the actinic rays! . . . You should be more careful, Tilly; you know the sun affects the spinal fluid and damages the ganglia and in the end will certainly send you mad.'

Piles of pith helmets receiving final inspection before being despatched to British troops, in 1942.

Cousin Hilary was not alone in his fear of the sun's sinister actinic rays. The British troops in India were commanded to wear a pith helmet or a solar topee at all times during the day and the punishment for not doing so was quite severe – confinement to barracks for fourteen days.

Not until 1917 was it established that sunstroke was a failure of temperature regulation rather than a direct effect of the tropical sun. The widespread belief in actinic rays took some time to die out, however, and was still considered a possibility as recently as 1927. Nowadays, sunstroke is referred to as heatstroke in recognition of its aetiology.

Exercise in a warm climate is a common cause of heatstroke. Risk factors include being unfit, not drinking enough in long events and a fast finishing pace. Amateur runners in marathon races are particularly susceptible and race organizers sometimes face the difficult decision of whether to cancel the race because the weather is simply too good. Nor are professional athletes exempt. In June 1999, Jim Courier collapsed suffering from dehydration and heat exhaustion after winning the second longest singles match ever played at Wimbledon – four hours and twenty-seven minutes. The many hundreds of spectators who watched the match were unaffected, for England in summer is rarely hot and it was not an unusually warm day. It was the endogenous body heat generated by the intensity of his epic contest that caused Courier's collapse.

Courier needed little more than intravenous fluids and rest. The movie star Martin Lawrence was less fortunate, for he spent three days in intensive care in a coma precipitated by heatstroke. Eager to lose weight for a new role, the actor went jogging in temperatures of 38 °C, clad in several layers of clothing. The summer heat of Los Angeles took its toll and he collapsed outside his home with a core temperature of 42 °C. He was very lucky to survive.

When evaporative cooling cannot take place, the risk of heatstroke is particularly high. Hot and humid environments not only feel awful, they are actually more hazardous. Evaporation of sweat may also be impaired by clothing. The new 'breathable' fabrics used for modern waterproof clothing allow sweat to escape and are far more comfortable to wear when walking than old-fashioned rubber rainwear. Impermeable garments can be dangerous if combined with intense exercise. One young British soldier taking part in a cross-country training run died from overheating because he was wearing a rubber diving suit. His clothing hampered moisture loss, so that the sweat pouring off him collected inside the suit and severely restricted his ability to lose heat by evaporation. Keen to prove his worth, he carried on, with tragic results.

In sedentary people, heatstroke is usually a result of impaired sweating. People with cystic fibrosis are particularly susceptible to heatstroke because they are unable to sweat. Although they normally tolerate the range of temperatures in the United Kingdom without difficulty, they may suffer from heat prostration in tropical climates. This sometimes happens when young British patients are taken for a trip to Disneyworld in Florida. Interestingly, the increased susceptibility of cystic fibrosis patients to heat stress was the key to understanding the basis of this disease. During a heatwave in New York in 1951, the paediatrician Paul di Sant' Agnese noticed that many children admitted to hospital with heat prostration also suffered from cystic fibrosis. Realizing the significance of this observation, he analysed their sweat and found that it contained an abnormally high level of salt. This finding forms the basis of the sweat test still used clinically to diagnose the disease. We now know that cystic fibrosis results from a genetic defect in a membrane protein that is involved in the transport of chloride ions out of cells. Sweat consists of a weak solution of sodium chloride (common salt), and the inability to excrete chloride hinders water flow into the sweat gland and so prevents sweat formation.

Even in ordinary people, prolonged exposure to a hot climate may result in a failure to sweat. This is usually preceded by an inflammation of the sweat glands, known as prickly heat. Prickly heat is characterized by itchy red 'little pimples rising over every inch of the body so that you couldn't put a pin between a pimple.' It affects one in three people exposed to hot climates, as well as those working in deep mines or with furnaces. In the days of the Raj, the British suffered terribly from prickly heat during the hot season in India. As one observer described,

> 'A chap would be playing cards and begin to scratch lightly. By the time the evening was out he'd be rushing round like a madman, tearing himself to pieces trying to quieten the irritation. I've seen one or two people rip themselves to ribbons, tearing their chests until all the skin was hanging down in layers.'

Although the inflammation eventually subsides, the sweat glands may cease to function, rendering the person susceptible to heatstroke. Fortunately, the condition is reversed on return to cool conditions.

Certain drugs can precipitate hyperthermia. Probably most notorious of these is Ecstasy,[3] a recreational drug often taken at 'raves' to give a 'high' or to maintain stamina while dancing. When combined with physical exercise, its use can trigger a potentially fatal rise in body temperature. So well recognized is the problem of Ecstasy-induced hyperthermia that some clubs provide special areas for 'chilling out' – a phrase that has now entered the English language.

Heatstroke occurs when the normal thermoregulatory system of the body fails and the core temperature increases to 41 °C or above. Its onset can be remarkably rapid. Early warning signs include a flushed face, hot dry skin, headache, dizziness, loss of energy and increased irascibility. Mental confusion and uncoordinated movements follow. Despite the elevated body temperature, sweating ceases so that the temperature may escalate even further. Death occurs when the temperature exceeds 42 °C.

Heatstroke is a medical emergency and requires immediate treatment. Untreated cases die of brain damage caused by the elevated body temperature and even with treatment the mortality rate is over 30

Quivering Pigs and Shivering Humans

About one in 20,000 people have a rare genetic disease called malignant hyperthermia. When these people are given common anaesthetic gases, such as halothane, their body temperature rises very rapidly, sometimes by as much as 1 °C every five minutes. This occurs because the anaesthetic triggers spontaneous contractions of the skeletal muscles – quite simply, victims shiver themselves hot. The disease is an anaesthetist's nightmare, for if it is not treated rapidly it can be fatal.

Muscle contraction is initiated by an increase in the intracellular concentration of calcium ions, which activates the contractile proteins. Calcium is normally locked away inside a special membrane-bound storage compartment in the muscle cell and released only in response to a nerve impulse. Patients with malignant hyperthermia have a defect in the protein pore that controls the release of calcium ions from their intracellular stores, and, in these people, anaesthetics open the gates of the pore allowing calcium to flood out of its store into the cell and trigger contraction. The muscle physiologist Shirley Bryant first suggested that the drug dantrolene, which blocks calcium release, might be an effective treatment for malignant hyperthermia and today this drug is kept in operating theatres throughout the world for just such an emergency.

Malignant hyperthermia is not confined to humans. It is also found in pigs, where it is known as porcine stress syndrome because it can be precipitated by stress (unlike the human disease). Exercise, sex, birth, the stress

per cent. The best way to cool a victim of heatstroke is to sponge them down with tepid water. Evaporative cooling lowers the skin temperature far more effectively than plunging them into a bath of cold water, which has the effect of causing a generalized vasoconstriction that directs the blood away from the skin and so restricts heat loss. In severe cases, ice-packs are applied to the body at places where large blood vessels come close to the surface of the skin, such as on the neck, under the arms and in the groin.

Fever!

Normally, the hypothalamic thermostat in man is set at about 37 °C, but during fever it may be reset to two to three degrees higher and the of transporting pigs to market, or even the conditions in which the animals are kept, can trigger a fatal rise in body temperature. It is of considerable economic importance, for when the afflicted pigs die of an attack their meat becomes tough and unsaleable. Until recently, porcine stress syndrome was widespread in Britain because pigs were selectively bred for lean meat, an attribute that turned out to be associated with malignant hyperthermia. This is because pigs with porcine stress syndrome have sub-threshold muscle activity (it is as if they were performing isometric exercises continuously) which leads to well-developed muscles and lean meat.

The pig has proved a very useful model for understanding the human disease. In 1991, the gene responsible for porcine stress syndrome was identified and shown to encode the muscle calcium release pore, and mutations in this protein were found that resulted in muscle contractions when the pig was exposed to halothane. All pigs with porcine stress syndrome have the same mutation in the gene, indicating that they are all descendants of a single founder animal that spontaneously developed the mutation some time in the past. The gene has now been bred out of the British pig population by identifying susceptible animals using a simple test. Young pigs are given a quick whiff of 3 per cent halothane to breathe. Those that carry the defective gene develop a transient muscle rigidity (from which they recover), and they are then removed from the breeding population.

Once the pig gene was cloned, it was relatively straightforward to obtain the human gene and show that it was responsible for malignant hyperthermia. A genetic test is now available for people suspected of having the disease.

temperature is then regulated, with the same sensitivity, about this new set-point. The resetting of the thermostat is due to the synthesis of chemical transmitters known as prostaglandins by the brain, triggered in response to bacteria, or to pyrogenic substances secreted by bacteria. Aspirin, which lowers fever, acts by blocking the synthesis of prostaglandins.

There has been considerable debate over the centuries about whether fever serves any beneficial function in infectious disease. One view, put forward by Thomas Sydenham in the seventeenth century, is that 'fever is a mighty engine which Nature brings into the world for the conquest of her enemies'. Rephrased in modern terminology, this argument suggests that fever is a natural part of the body's defence against infection which has arisen because some bacteria are more susceptible to an elevated temperature than our own cells. The alternative hypothesis contends that fever is merely symptomatic of the severity of infection and that far from having any therapeutic value, it may actually compromise the ability of the patient to fight infection. This debate is not merely of academic interest. At its heart is the question of whether or not an effort should be made to reduce the patient's temperature to its normal value.

The issue is still not fully resolved and there is evidence to support both sides of the argument. Nevertheless, most people favour the view that a rise of 1–2 °C in body temperature is not deleterious and may even be beneficial in adults.[4] This idea is supported by the fact that the survival rate of lizards suffering from bacterial infection is markedly greater when they are placed in a warm environment than in a cold one. Because the body temperature of the lizard equilibrates with that of its surroundings, this finding argues that an elevated body temperature enhances the ability to fight infection. Indeed, before the advent of antibiotics, fever therapy was successfully used as a treatment for gonorrhoea and syphilis. Artificial fever was induced in several ways – most dramatically by infection with the malaria parasite, which was later killed by quinine. If the person survived this ordeal, the syphilis bacterium was sometimes killed – with luck, before its human host – and the patient cured. An unusual form of trial by fire, perhaps.

Life Without Water

Just as food is the key to survival in the cold, so water is the limiting factor for life in the hot zone. The ability to cool off by sweating copiously is dependent on the availability of water and the chief difficulty with life in the desert is not that it is hot, but that it is arid. While people can go for many days without food, they do not survive long without water. Hunger-strikers, it may be noted, never refuse water, presumably because their demise would be too rapid to make sufficient public impact.

When water lost in sweat is not replaced by drinking, dehydration occurs. This stimulates the secretion of hormones that act both to conserve water, by reducing the amount of water lost in the urine, and to increase water intake, by making you feel thirsty. If water loss continues, the face and eyes take on a gaunt sunken look. There is also an associated weight loss, a fact which is exploited by jockeys and boxers struggling to make weight limits, who 'sweat off' the extra pounds in the sauna. Most people can tolerate a 3–4 per cent decrease in body water without difficulty. Fatigue and dizziness occur when 5–8 per cent is lost, while over 10 per cent causes physical and mental deterioration, accompanied by severe thirst. Loss of more than 15–25 per cent of the body weight is invariably fatal.

Although a person will die when their body water is reduced by 15 per cent, the camel is unaffected until as much as 25 per cent of its body water has been lost, which enables it to go for seven days without food or water. One reason for the camel's remarkable ability to withstand dehydration is that it is able to prevent a fall in blood volume, despite significant water loss. Even when a camel has lost a quarter of its body

Measurement of oxygen consumption in a camel. The camel is wonderfully adapted to desert life. Its heavy fur acts as a good insulator that decreases heat gain and its long slim legs provide a large surface area for heat loss. When water is scarce, it allows its body temperature to rise by as much as 6°C before it starts sweating, thereby conserving water. The stored heat is lost at night, when the air temperature is cooler, without evaporative water loss. Heat storage during the day not only reduces water loss, it also decreases the heat gradient between the environment and the body surface, and so reduces heat uptake. The camel tolerates considerable water depletion without apparent harm and its hump acts as a fat store for when food is scarce. Its ears and nostrils are lined with fine hairs to filter out the dust of the desert, and it has a double row of long eyelashes that serve the same purpose.

water, the blood volume decreases by less than a tenth. In contrast, the blood volume in humans drops by about a third, increasing the viscosity of the blood. The thicker blood circulates more slowly and is harder to pump, so that less heat is lost through the skin and the person suffers a fatal rise in body temperature. It also enhances the risk of a stroke. Water deficiency not only decreases the volume of the blood and extracellular fluids; water is also sucked out of the cells, causing them to shrink and damaging the cell membrane and cellular proteins.

Death by dehydration is not easy, for the victim is tormented continuously by a burning thirst. With remarkable fortitude, several people have recorded the experience for posterity. One of these was Antonio Viterbi, a judge advocate under the first French Republic, who was sentenced to death by the court of Bastia (in Corsica) during the Restoration for his political beliefs. To avoid the shame of the scaffold, he elected to die by depriving himself of food and water. It took seventeen painful days and quite extraordinary will-power. His diaries reveal that whereas hunger forsook him after a few days, he endured an unremitting and intolerable thirst.

When lack of water is accompanied by intense heat, dehydration and death occur much faster than in the case of Viterbi – almost half the victims die within the first thirty-six hours. Tales of desert travellers who ran out of water with fatal, or near fatal, consequences are legion. As one experienced desert traveller remarked:

> 'In that terrific temperature, one's bodily moisture must be constantly renewed, for moisture is as vital as air. One feels as though one were in the focus of a burning glass. The throat parches and seems to be closing. The eyeballs burn as though facing a scorching fire. The tongue and lips grow thick, crack and blacken.'

One of the most famous, and most remarkable, cases of desert survival is that of the Mexican Pablo Valencia, who lost his way in the Tinajas Atlas region of south-western Arizona in the summer of 1905. He spent seven days and nights without any water in temperatures that ranged between 85 °F by night and 95 °F by day. When he was found, he was stark naked and burned black by the sun. His formerly well-muscled legs and arms were shrunken and shrivelled, 'his lips had disappeared as if amputated, leaving low edges of blackened tissue . . . his eyes were set in a winkless stare . . . and he was deaf to all but loud sounds, and so blind as to distinguish nothing save light and dark.' He was also unable to speak or swallow because his mouth was so parched. But Pablo was lucky, for he did not suffer from the delirium or wild movements (resembling an epileptic fit) that sometimes afflict those suffering from extreme dehydration and speed their demise. He was able to stagger on, slowly and painfully, and to recognize a familiar landmark when he encountered it. In fact, he had almost made it back to camp before he was found. His rescuers doused him with water, and forced water and dilute whisky between his lips. Within an hour he was able to swallow, within a day to speak, by the third day to see and hear, and in a week he was well and cheerful, and had gained 8 kilos in weight.

Not everyone has the iron constitution of Pablo Valencia and death from a combination of dehydration and heatstroke can be far more rapid. Lowell and Diana Lindsay recount how an experienced desert motorcyclist felt faint on a hot afternoon trip to the Anza desert, and sent his party on ahead to seek help. Left alone, he became dehydrated and delirious, and disregarding the plan to sit still and await help, set off after them down the aptly named Arroya Seco del Diabilo (the Dry Wash of the Devil). Four hours later, his body was found by a ranger. Sadly, such stories are not uncommon, even today. Nor do they necessarily occur in remote areas: a breakdown on a desert road, or losing the trail on a day hike, can have fatal consequences if you run out of water.

When very active, you lose more water than you consume voluntarily. You simply do not drink enough to prevent dehydration and it is possible to be incapacitated by lack of water without feeling unbearably thirsty.[5] Only when rested and fed, do you drink sufficient water to replace that lost in sweat. If exercising in a hot climate, it is therefore necessary to drink even when you do not feel the urge. If water is short, however, your best strategy is to cease activity and sit quietly in the shade. Exercise will simply make you hotter, so walk at night when the air is cooler. Do not try to make your water last longer by not drinking it: it is imperative to drink when you feel thirsty and one of the maxims of desert life is to 'store water in the body rather than in your bottle'. The camel exemplifies this principle to perfection. It can drink up to 120 litres on reaching water, in as little as ten minutes. In contrast, severely dehydrated humans must sip water sparingly, despite their raging thirst. Too profligate a consumption after fasting only results in vomiting.

Desert creatures must make use of every drop of water that they find and they have evolved some extraordinary ways of doing so.

Desert beetles collect water by condensation, lining up on the crest of a sand dune and orientating their body to the cool morning breeze. The breast feathers of the male sandgrouse soak up moisture like blotting paper, so after drinking its fill the bird immerses its breast in water, loading up its feathers before flying back to its chicks. This enables the sandgrouse to breed deep in the desert, many miles from water. The water-holding toad of the Australian outback stores water in its bladder when times are plentiful and builds a watertight underground chamber where it can live for several years when water is scarce. This provides a useful emergency water supply for Aborigine people.

Mammals have also evolved specialized ways of reducing water loss. Kangaroo rats possess a specialized heat exchanger in their nasal passages that cools the exhaled air to a temperature below that of the body, causing the water vapour it contains to condense on the nasal passages. This reduces evaporative losses. Some birds employ a similar strategy. Humans do not possess this ability and lose water continuously from their respiratory surfaces (a great deal of water is lost through the lungs).

Food metabolism produces waste products, such as urea, that require water for their elimination. Humans cannot produce a very concentrated urine. Some desert animals can do much better. Many never drink during their brief lives but obtain all their water from their food. These animals have extremely efficient kidneys that produce a highly concentrated urine, enabling them to use as little as a quarter of the water that the human kidney needs to excrete the same amount of urea. Birds go one better still. They excrete uric acid, which requires very little water for its elimination. The resulting solid or semi-solid white waste is familiar to anyone who has stood in a gull colony or a flock of pigeons.

Oceans are also deserts for, as is well known, it is not possible to drink sea water for survival. Sea water contains a higher salt concentration than can be excreted by the kidneys, so drinking sea water will actually accelerate dehydration. Marooned on a liferaft in the middle of the ocean with a hot sun beating down, your best bet is to douse yourself with sea water to facilitate evaporative cooling. This helps conserve water by reducing the need for sweating. For similar reasons, when John Fairfax rowed single-handed non-stop across the Atlantic Ocean in 1969 he slept during the heat of the day and rowed at night under the stars, when it was cooler.

Salt of the Earth

Sweat contains a significant amount of salt. The more you sweat, the more salt will be lost, and in a hot environment this can be quite considerable – up to 12 grams a day or almost three teaspoons of salt. The body copes with this problem by secreting a hormone that enhances salt conservation by the kidneys so that less is lost in the urine. It also stimulates the taste for salt, so that you eat more of it.

My grandfather worked as a foreman in a factory making locomotive wheels. Molten steel flowed from giant furnaces into huge open cauldrons which carried the white-hot metal to another part of the factory where it was poured into the wheel moulds. The intense heat meant that the workers sweated profusely, rendering them prone to salt and water deficiency. My mother was fascinated by her father's passion for salt sandwiches – a seemingly inexplicable taste to a young child, but one that has a sound physiological basis, for the salt lost in sweating must be replaced in the diet.

Salt depletion results in painful muscle cramps in the arms and legs, which are sometimes known as 'stokers' cramps' because they were common among those who shovelled fuel into ships' engines. Others who work under hot conditions, such as miners or athletes exercising in hot environments, may also suffer from them. Cramps only develop when salt depletion is associated with muscular exercise. In less active people, salt depletion causes fatigue, lethargy, headache and nausea. The treatment is to eat more salt – one of the few times when this is actually recommended by the medical profession.

A Hot Cradle for Mankind

Sweating is the key to survival in the hot zone. Humans can tolerate considerable dry heat if they are supplied with plenty of water (and salt) to replace that lost in sweat – it is not so much the high temperature as the scarcity of water and shade that makes the desert a dangerous place. When intense heat is associated with high humidity, however, evaporative cooling is no longer possible and the risk of heatstroke rises dramatically. Physiologically, humans are poorly adapted to such conditions and our survival in hot and humid environments depends on a combination of behavioural adaptations and the use of technology (air-conditioning, for example). By contrast, we are rather well adapted to dry heat, for we are extremely well endowed with sweat glands, which gives us one of the highest sweat rates of any mammal, we are almost hairless, and we have relatively long slim limbs. This suggests that humans evolved in a hot environment, where the greatest problem was to lose heat rather than conserve it. Our physiology lends support to the fossil evidence that *Homo sapiens* originated on the hot plains of Africa.

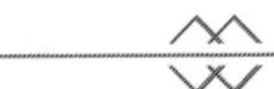

The magnificent frozen landscape of Antarctica, the most pristine wilderness on Earth.

Cold Water Blues

IT WAS A bitterly cold Easter afternoon. We had spent the last week sailing around the Inner Hebridean islands and were now safely back at anchor in Dunstaffnage Bay, close to Oban, getting ready to go ashore. We were glad to leave, for a storm was brewing. Heavy grey clouds massed above us, turning the sea the colour of lead. The wind screamed through the rigging and tugged at my clothing. It lifted the sea into a heavy chop, and blew thin sheets of spume from the tops of the waves, enveloping us in a fine mist of freezing foam. It pushed fretfully against the bow of the boat, causing it to veer and snatch at the anchor. The tide was ebbing rapidly and the boat surged erratically through the water, caught now by the wind, now by the tide. I stumbled about the bucking deck gathering together our belongings. Attracted by an unexpected sound, I looked up from my task to see a small rowing boat making its way towards the yacht on the adjacent mooring. Its occupant, a middle-aged man, struggled to make headway against the wind and tide. When he arrived at his target, the two boats dipped and bowed to each other in a complicated dance that made it difficult for him to bring the smaller boat alongside. He made a wild grab for the anchor chain and missed. This was a mistake, for the lightweight fibreglass dinghy reacted to the sudden shift in weight, tilting sideways and sliding him into the sea in one slow easy movement, like an egg slipping out of its half-shell. The dinghy filled with water and sank instantly, while the tide caught the man, spun him around and sucked him out to sea.

My companion, Tim, reacted more quickly than I did. Shouting to me to pass him a rope and make fast the other end to the winch, he leapt into the rubber dinghy that was tied to the back of our boat and rowed energetically towards the distant figure. It seemed to take an agonizingly long time before Tim reached his goal, yet it was probably only a few minutes. Nevertheless, by the time he arrived, the victim was so cold that he was barely able to move, his teeth were chattering so much that his speech was incomprehensible and his hands and fingers had lost their ability to grip. Wearing no life-jacket, and in a heavy chop, he kept sinking below the waves and was far too exhausted and disorientated to assist in climbing into the boat. For his rescuer, it was like lifting the proverbial sack of potatoes – and a heavily waterlogged one at that – and the difficulty was exacerbated by the tide which carried the dinghy out to sea until the line over which I stood guard snapped taut, with the dinghy dancing at the end of it. The weight of the semi-conscious man bore down heavily on the side of the dinghy, threatening to capsize it. It took many minutes before Tim managed to drag him into the dinghy and a considerably longer time to rewarm him (and his rescuer) subsequently. But the victim was fortunate. Immersion in cold water can kill rapidly, the bay was deserted apart from ourselves and the tide was strong. Once outside the bay in the open sea, there was little chance he would have survived.

Later that night, reflecting on the events of the day, I thought of my grandfather, Walter Blackburn, and of how without his even more miraculous escape from hypothermia, I would not be here to tell his story.

At the outbreak of the First World War in 1914, Walter was twenty-three years old. Young and full of idealism, eager to serve his country and quite unaware of what awaited him in the trenches, he enlisted early. He was assigned to the Royal Armoured Medical Corps as a medical orderly, where he was rapidly initiated into the harsh realities of warfare. While assisting at his first operation in a tense, overcrowded tent, he was tossed an amputated leg by a harassed surgeon and brusquely told, 'Here, boy, get rid of this.' He fainted with shock.

Within months of his arrival at the front, Walter was shot in the knee and incapacitated. Worse still, the wound became infected. In those pre-antibiotic days, no treatment was available and the septicaemia spread. He was evacuated to England in a critical condition. The ship that carried him across the Channel was so heavily loaded with casualties that there was no room for them all below deck, and those deemed less likely to survive, who included Walter, were left on deck exposed to the cold, wind and rain. Despite sailing under the cover of darkness, their ship was torpedoed and sank. As it settled in the water, the deck tilted and my grandfather, strapped tightly to his stretcher and half-delirious with fever, slid gently into the sea. Being made of wood and canvas, the stretcher floated. It is not certain how long Walter lay in the cold water before he was eventually rescued, but it is clear that it was many hours.

Several things probably contributed to his remarkable survival. First, he was unable to move, for straps across his arms and legs pinned him to the stretcher. Consequently, his rate of convective heat loss was reduced, because the thin layer of water adjacent to his body was not dissipated by struggling. Second, he was a large man with a reasonable insulating layer of subcutaneous fat. Third, the fever elevated his metabolic rate, increasing his rate of heat production. Whatever the reason, he was fortunate because many people, including able-bodied men, lost their lives to hypothermia that night.

My mother possesses a poignant memento of his ordeal. As Walter lay semi-conscious in the water, he glimpsed a small blue handkerchief floating close beside him. Later he would say he felt that if he could only hold on to this scrap of blue silk he would survive, and when he was rescued, his fingers were still curled around the ragged blue charm.

4

LIFE IN THE COLD

'And now there came both mist and snow.
And it grew wondrous cold:
And ice, mast high, came floating by,
As green as emerald.'

SAMUEL TAYLOR COLERIDGE, 'The Rime of the Ancient Mariner'

COLD IS FOUND at altitude, at the poles and in the ocean depths, and as much as half the Earth and one-tenth of the oceans are covered with snow and ice for part of the year. Winter turns our landscapes into a frozen paradise of magical beauty but our bodies are ill-equipped to deal with cold and it can be lethal. Most animals, including man, abhor cold. So it comes as no surprise that when Dante wrote the *Inferno* he placed the circles of ice below those of fire.

The coldest temperature ever recorded on Earth is -89°C (-129°F), and was measured on 21 July 1983 at the Russian research station at Vostok on the Antarctic icecap. Even this is relatively warm when compared with the outer planets – Pluto's surface temperature is a glacial -220°C. Although not as cold as Antarctica, severe weather is regularly encountered in the Arctic circle, in mountainous areas and in other places where people normally live. In Siberia, for example, the temperature often dips below -60°C in winter. Britain is far more fortunate. Braemar has the distinction of being its coldest town, at a mere -27°C.

The air temperature decreases by 1°C for every 100 metres you ascend, so the high mountain peaks are always swathed in snow and ice and vie with the polar regions for the coldest places on Earth. On the summit of Everest it is regularly below -40°C, and biting winds lower the apparent temperature further still. The oceans are far less cold than the landmasses. Throughout much of the ocean depths the temperature is a chilly but constant 2°C, although the surface waters of Antarctica may fall to -2°C before they freeze because the high concentration of dissolved salts lowers the freezing point.

Fighting Cold

Every year, millions of people throughout the world experience weather conditions that put them at risk of cold injury. Humans can withstand considerable cold as long as they are well wrapped up, well nourished and have adequate shelter, so in peacetime people rarely sustain serious consequences, unless they happen to be victims of natural disasters such as earthquakes and avalanches. Polar explorers, mountaineers and skiers may suffer if they are caught without sufficient food or shelter, as may those who swim in cold waters, either by accident or design. But, in general, casualties in peacetime are few.

War is a very different matter. Cold has had a devastating effect on military campaigns and in doing so has shaped the course of history. Of the 90,000 infantry, 12,000 cavalry and forty famous war elephants that set out with Hannibal in 218 BC to march across the Alps, only about half reached northern Italy. The rest died from exposure *en route.* In 1812, with an army of over half a million men, Napoleon marched on Moscow. The countryside, stripped by the retreating Russian army, could not support the vast number of invaders and famine killed many thousands. Winter, Russia's traditional ally, arrived to complete the catastrophe. The temperature plummeted to more than 40 degrees of frost, furious winds whipped up thick whirlwinds of snow, ice enveloped the army, and thousands more men died. Less than 20,000 returned. As one survivor said, 'the army was swathed in an immense winding sheet' of snow. Hitler did not heed Napoleon's experience, for he too lost many thousands of men to the Russian winter during the Second World War. In November and December 1941, 10 per cent of the German army (about 100,000 men) sustained debilitating cold injuries, necessitating 15,000 amputations. Considerable casualties also occurred as a result of cold weather in the American Civil War, the trenches of the First World War, the Korean War, and the Falklands campaign.

Refugee populations may suffer severely from cold, for they often lack both shelter and food. The misery of the thousands of ethnic Albanians fleeing Kosovo in the spring of 1999 was compounded by

Ranulph Fiennes and Mike Stroud on their epic journey across Antarctica

freezing rain, and many elderly people and young children forced to sleep outside at night died of hypothermia.

A key lesson from military campaigns and polar expeditions is that hunger and hypothermia march hand in hand. Our bodies can only generate enough heat to keep us warm when provided with sufficient food, and the increase in calorie requirements produced by severe cold can be dramatic. In 1991, Sir Ranulph Fiennes and Dr Mike Stroud crossed Antarctica on foot, hauling all their food with them. Stroud, a doctor with a keen interest in physiology, calculated they would need 6500 calories a day. Because this would make the sledges unacceptably heavy, they compromised on 5500 calories a day and a consequent loss of body weight. Even then, they set out pulling 220 kilos (480 pounds) each.[1] Their journey was far more arduous than anticipated, for the temperature was so low that the sledges did not run smoothly over the fractured ice. Sledges and skates glide on a thin film of water, produced by the melting of the ice under pressure, but in the glacial conditions encountered by Fiennes and Stroud it was too cold for ice to melt under the runners and the sledges stuck as if in sand. Terrible winds and white-outs also hampered their progress. By the time they reached the South Pole they were emaciated, sick, very hungry and had each lost more than

20 kilos – equivalent to an energy expenditure of over 7000 calories a day. Stroud calculated that on one day he had used an astonishing 11,650 calories, the highest energy expenditure ever recorded in man. In contrast, most people lead such sedentary lives that a daily intake of 2500 calories for men, or 2000 for women, is more than sufficient.

How Cold Can You Go?

The lowest temperature that humans can tolerate is determined by the duration and extent of exposure. This means that, unlike most of the other extremes discussed in this book, it is not easy to give a clear limit. A nude person feels cold when the ambient temperature falls below about 25 °C and physiological responses are triggered if they do not take evasive action (by putting on some clothes or turning up the heat). These responses enable well-fed adults to maintain their core temperature in still air of between zero and +5 °C wearing only light clothing. In colder air, or if heat loss is increased by wind, rain or immersion in cold water, the body temperature will fall and hypothermia eventually occurs. Providing you are well wrapped up, however, it is possible to endure extreme cold. But local cooling of the extremities below -0.5 °C, the freezing point of human tissues, must be avoided.

It is well known that wind exacerbates the effect of cold. The term 'wind-chill factor' was coined by the American explorer Paul Siple to describe the fact that wind enhances the rate of heat loss (because it removes the surface layer of warm air and replaces it with cold). While visiting Antarctica in 1941, he and Charles Passel carried out a series of simple but ingenious experiments in which they compared the time it took for baked-bean tins filled with water to freeze at different temperatures in either still air or when exposed to strong winds. They found a marked difference in the freezing rate and subsequently devised a formula that enables the cooling power of the wind to be estimated in terms of a 'wind-chill equivalent temperature'.

In still air of -29 °C, there is little danger for a properly clothed person. If the wind is a mere 10 miles per hour, however, the temperature drops to the equivalent of -44 °C and skin freezes within one or two minutes. Increase the wind speed to 25 mph and the equivalent temperature is -66 °C. There is then severe danger, with flesh freezing within thirty seconds. The wind-chill factor can mean that even in air temperatures of around zero, frostbite of the extremities is possible. However, the widespread use of Siple's formula to calculate wind-chill factors can sometimes be overly dramatic when applied to humans, because we tend to wear more clothes on windy days and only our extremities behave like baked-bean tins.

To prevent the skin freezing in cold air, it needs to be supplied with plenty of warm blood. Unfortunately, this has the obvious disadvantage that more heat will be lost to the environment and the body will cool overall. There is therefore a trade-off between losing body heat and freezing of the peripheral tissues. Heat loss from the hands, feet, nose, ears and so on, is particularly high because of their large surface-area-to-volume ratio, and if the ambient temperature is very low, these are sacrificed and allowed to freeze in order to keep the central body warm. While losing a few fingers to frostbite is not pleasant, it increases your chance of survival.

In intense cold, even a good blood supply cannot prevent the skin from freezing. At -50°C, for example, bare skin freezes within a minute. Icy winds blowing on the face may freeze the surface layers of the eye, as sometimes happens to downhill skiers who forget to wear goggles. Eyelashes freeze together, glueing the eyes shut and lumps of frozen breath collect on men's beards forming a tinkling necklet of icicles. And in extreme cold, the breath turns to crackling ice crystals when exhaled, a phenomenon that has been called by the magical name of the 'whispering of the stars'.

While most of the body surface can be protected by clothing, the lungs are unavoidably exposed to the freezing air. Because the air is warmed as it enters the respiratory tract they are not normally injured by cold. If the air is both very cold and dry, however, the cells that line the respiratory tract may be destroyed and sloughed off, as graphically described by T.H. Somervell, a surgeon who nearly suffocated from the phenomenon while climbing Everest in 1936:

> 'When darkness was gathering I had one of my fits of coughing and dislodged something in my throat which stuck so that I could neither breathe in nor out. I could not of course make a sign to Norton or stop him, for the rope was off now, so I sat down in the snow to die while he walked on. I made one or two attempts to breathe but nothing happened. Finally, I pressed my chest with both hands, gave one last almighty push – and the obstruction came up.'

Freezing of the lungs has also been described in horses and in sledge dogs, but has never been reported for people working in Antarctica.

Rapid freezing can occur if bare skin contacts metal, because it is such an excellent conductor of heat. During the Second World War, terrible frostbite injuries were sustained by the 'waist-gunners' of B-17 and B-24 bombers of the United States. Their planes flew at altitudes of 7600 to 10,700 metres, where the outside air temperature was -30 to -40 °C. To operate the machine-guns, the gunners had to open the 'waistports' to the outside, so that the freezing air rushed in and swirled around inside the plane. Often, the gunners abandoned their mittens, which they felt hampered their dexterity, and worked the machine-guns bare-handed. Contact with the bare metal, combined with lack of oxygen, fear and exhaustion, exacerbated the effects of cold and despite lasting only for a minute or two, such encounters resulted in severe frostbite.

In such extreme cold, bare hands stick fast to metal because the moisture on the skin freezes on contact with the metal, and pulling the hand away can leave a layer of skin behind. My schoolfriend's father was a doctor on an Antarctic expedition and I well remember the day we found his briefing notes in their attic. They advised that if by chance your skin became frozen to exposed metal, you should urinate on the affected part, as the warm urine would melt the ice and release the hand unharmed. It was a clear reminder that their expedition – like so many others before them – had been an all-male one, for the difficulty that a woman would have in following these instructions is only too obvious.

Water saps body heat far more rapidly than air, and survival times in water are far lower than those in air of the same temperature. Because the cold seas around Antarctica may not freeze until the water temperature falls below -2 °C (due to their high salt concentration), placing an unprotected hand in the water can even result in frostbite. Fish that live in Antarctic waters have substances in their blood that act as natural antifreezes and prevent ice crystal formation.

Even mild cold has obvious effects on the body. It impairs nerve function and decreases sensation and manual dexterity. The difficulty you experience on a cold frosty morning fastening the buttons of your coat is due to the slowing of the nerve signals from your brain to your fingers. Cold muscles also work less fast, making the fingers stiff and clumsy. The critical air temperature for manual dexterity is 12 °C and that for touch sensitivity is 8 °C. Low temperatures also impair the function of the sensory nerves that transmit pain, which is why putting an icepack on a sprained ankle or a burn helps relieve the hurt.

The anaesthetic properties of cold were exploited by soldiers during the great retreat from Moscow in 1812, who used their horses as a living larder. It was too cold to butcher them for food, for the men's hands were too numb and the carcasses would have frozen like iron. So, in the words of August Thirion, senior sergeant of the 2nd Cuirassiers:

> 'We cut a slice from the quarters of horses still on their feet and walking, and these wretched animals gave not the least sign of pain, proving beyond doubt the degree of numbness and insensitivity caused by the extreme cold. Under any other conditions these slices of flesh would have brought on a haemorrhage and death, but this did not occur with 28 degrees of frost. The blood froze instantly, and this congealed blood arrested the flow. We saw some of these poor horses walking for several days with large pieces of flesh cut away from both thighs.'

Some people have a rare inherited disease called paramyotonia congenita that renders their muscles peculiarly sensitive to cold and causes them to become paralysed when the temperature drops. They often discover their problem during cold weather, when their hands become clamped to the cold metal handlebars of their bicycle, they are unable to release their grip on the spade they have been using to shovel snow, or they become stiff and weak after playing football. Other people realize it when they develop a stiff tongue and slurred speech after eating ice cream or iced drinks. Patients with paramyotonia congenita have a mutation in the gene that codes for a protein known as a sodium channel, which is important for the conduction of electrical signals along muscle fibres. These signals are essential for the initiation of muscle contraction and in their absence the unfortunate individual is paralysed. The condition is not life-threatening (it does not paralyse the respiratory muscles) but it is undoubtedly very inconvenient.

Most people will be well aware that exposure to cold has another effect: it increases urine production. This is because urine output is related to the volume of circulating body fluids and any increase in volume is sensed by pressure receptors and stimulates urine production. When the surface blood vessels constrict due to cold, the volume of the circulatory system is reduced and thus the blood pressure increases. At very low temperatures, the ability of the kidneys to produce a concentrated urine also fails. Dehydration caused by excessive fluid loss is a significant problem for people, like mountaineers, who are exposed to cold for long periods.

Life in the cold also poses many practical difficulties. At -55°C, planes can't fly; fuel and radiators freeze and must be unthawed before trucks can start; and batteries do not hold their charge. In much of Canada, cars that are parked outside during a wintry day must be plugged into an electricity supply to ensure that their owners can drive home after work. The limitations of modern transport in temperatures well below freezing were vividly illustrated in the winter of 1998, when many Siberian villages that are accessible only during the brief summer months were discovered to have received insufficient food and fuel to carry them through the winter. Even with modern technology, they were very hard to reach and the population suffered from hunger and cold. In intense cold, synthetic fabrics may tear, so that furs become a necessity. Metal power lines stretch and snap under the weight of icicles, cutting off electrical power supplies. Without an alcohol thermometer it is not even possible to measure how cold it is, for mercury freezes at -39°C. On the bonus side, low environmental temperatures mean that milk can be sold in handy frozen blocks and food can simply be stored on the porch, so that refrigerators become superfluous.

Coping with Cold

Venture outside in shorts and a thin T-shirt on a snowy winter's day and the cold takes your breath away. Your skin blanches, goose pimples cover your bare arms and you start to shiver violently as your body reacts to the cold by reducing heat loss and increasing heat production.

Heat flows from hot to cold objects so that all animals that maintain their body temperature higher than that of the environment, like humans, will lose heat constantly. As explained more fully in Chapter Three, the rate of heat loss is determined by the amount of warm blood flowing near the surface of the skin, and the greater the blood flow, the more heat will be lost. A key strategy for conserving heat is therefore to reduce the flow of blood to the skin. However, this can only be tolerated without injury to a limited extent, because the surface tissues will be deprived of oxygen and nutrients.

When the air temperature falls, the blood vessels in the skin constrict, shunting warm blood away from the surface, so that the skin turns pale and heat is conserved. Paradoxically, when the temperature drops below about 10 °C, the surface blood vessels in the skin dilate rather than constrict and if the temperature falls even further, periods of vasodilation alternate with periods of vasoconstriction. These oscillations prevent the skin being damaged by severe cold and ensure that it receives an adequate, if intermittent, oxygen supply. The phenomenon explains the red nose and hands characteristic of frosty weather and is particularly well developed in those who work outdoors in cold climates, like fishermen. You can test this response quite easily for yourself by immersing one hand in cold water. At first, the reduction in temperature stimulates the blood vessels to constrict and your skin will turn white. Gradually, your hand will start to hurt and become increasingly painful. This is probably the result of the build-up of toxic metabolites caused by the lack of blood flow. After about five to ten minutes, however, the skin will turn red and the pain simultaneously ease as vasodilation occurs.

If it is very cold, the loss of heat from the skin can be excessive even if the blood supply is only intermittent. Under these conditions, the surface blood vessels constrict continuously. The bloodless regions then cool to the environmental temperature and frostbite may occur.

Goose pimples are an obvious sign that someone feels chilly. They are caused by contraction of the muscles surrounding the hair follicles, a phenomenon that goes by the wonderful name of horripilation. In man it serves no useful function as our hairs are too few for their erection to have any insulating function. In other mammals it is quite a different story, as described later.

In addition to reducing heat loss, the body reacts to cold by increasing heat production. The most important source of heat in adult humans is muscle activity, for muscle contraction is inherently inefficient and heat is produced as a by-product. Stored chemical energy is thereby converted into heat. When the sun slides behind a cloud on a cool summer afternoon, we start to shiver. Shivering results from involuntary contractions that cause the muscles to quiver. It begins in the muscles of the trunk and arms but ultimately progresses to the jaw muscles, making the teeth 'chatter' and shaking the body with large shudders.

Shivering can increase heat production five-hold but its efficiency is partially offset by the enhanced convective heat loss to the environment that results from trembling. This may diminish its usefulness, particularly in children, where the surface-area-to-volume ratio is greater. Voluntary exercise also generates large amounts of heat. We all know that jumping up and down, stamping our feet and slapping our arms makes us feel warmer – just watch the spectators at a football match. Heat production, whether from shivering or voluntary exercise, is limited by the size of the body's fuel store. Thus how long and how effectively we shiver is determined by the amount of glycogen stored in our muscles: usually a few hours is the limit. Physical fitness, stamina, and fuel reserves also limit voluntary exercise. Ultimately then, heat production depends on the available supply of food.

Babies have a proportionately greater surface-area-to-volume ratio than adults and therefore lose heat more readily. They are extremely sensitive to cold, but they do not shiver. Instead, they possess a specialized heat-generating system. Fat pads are found along the shoulders and upper backs of babies, and around their kidneys, and amount to around 4 per cent of the total body weight. This fat differs from normal fat stores and is known as brown fat. While white adipose tissue serves as an insulating blanket, brown fat is more like an electric blanket. Its characteristic brown colour results from the fact that the cells are loaded with large numbers of pigmented organelles called mitochondria. Normally, mitochondria act as biochemical furnaces that burn fuel to produce chemical energy – hence they are widely referred to as the powerhouses of the cell. Brown fat mitochondria, however, burn fuel to produce heat. This is achieved by a specialized protein that uncouples fuel metabolism from energy production. The uncoupling protein also regulates energy balance and protects against cold in some adult animals; mice which lack this protein are far more cold-sensitive than normal mice.

Heat production in brown fat is stimulated by the stress hormone adrenaline. An extensive network of fine blood vessels ramifies through the tissue and carries the heat away to warm the rest of the animal. In humans, brown fat is mostly confined to babies and by the time the infant has reached adulthood it has almost vanished. Only a few brown fat cells remain in adult life, dispersed throughout the white fat. The brown fat is retained, however, by many small mammals, particularly hibernators such as bats, ground squirrels, hedgehogs and marmots,

Pads of brown fat over the shoulder blades and encircling the neck provide newborn babies with a built-in heater

that use the heat generated by brown fat to rewarm their bodies when they awake from hibernation.

Parenthetically, one might imagine that increasing uncoupling protein activity would be a very effective means of enhancing weight loss. Indeed, it has been suggested that thin people have a higher basal activity of their uncoupling proteins, so that they burn away calories as heat rather than storing them as fat. This might explain why two people can eat just as much as each other, but only one of them becomes fat. Although adult humans lack significant amounts of brown fat, recent work suggests that related uncoupling proteins are present in other tissues. Once piece of evidence for this idea is that mutant mice which lack uncoupling protein in their brown fat cells are cold-sensitive but they do not get obese. This implies there may be an additional type of uncoupling protein that aids body weight regulation.

Another specialized form of biological heater is found in swordfish and billfish (a group of oceanic fish that includes the blue and black marlins much prized by sports fishermen for their acrobatic leaps when caught on rod and line). The heat-producing tissue is a modified eye muscle that lies below the brain and keeps the eyes and brain at a relatively constant 28 °C, while the temperature of the rest of the body is allowed to fluctuate with that of the surrounding water. During a deep winter dive, this can fall as low as 8 °C. Most of the blood entering the brain is diverted through the heater organ and warmed during its transit. Before entering the heater organ, the blood vessels break up into an extensive network of fine arteries which lie in close apposition to the small veins leaving the brain. Thus the warm blood leaving the brain transfers its heat to the cold blood entering, and so reduces the work of the heater tissue. The heat-producing organ is composed of modified muscle cells, which have almost no contractile tissue and are packed full of mitochondria. Unlike brown fat, uncoupling of energy production from fuel metabolism does not appear to take place. Instead, the energy (ATP) produced by metabolism is immediately consumed in futile biochemical cycles that serve no function other than to produce heat as a by-product.

Surprising as it may seem, plants also regulate their temperature by increasing heat production. A common sight in the hedgerows of the Dorset countryside where I grew up was the wild arum lily *Arum maculatum*, known locally as cuckoo pint, Lords and Ladies, or Jack-in-the-pulpit. The latter name may have been a reference to the phallic-looking purple spadix of the plant, which carries the pollen. The spadix generates so much heat that it volatizes chemicals, producing a powerful smell that attracts flies and other insects to fertilize the plant. The heat production can be very significant, raising the temperature to as much as 45°C. It is amazing that the plant does not cook itself, but the spadix is apparently adapted to the heat. The alpine snowbell (*Soldanella montana*) is even more extraordinary. This attractive little plant generates sufficient heat to melt the surrounding snow and so preserves itself from frostbite.

Frozen to Death

Every year, mountain rescue teams all over the world are called out to rescue people caught in unexpected snowstorms, buried in avalanches, or lost or immobilized by injury without adequate clothing. Most of them will be suffering from the effects of cold.

The oldest known case of hypothermia is probably that of Ötzi, a herdsman who died over 5200 years ago in the Ötz valley, which lies high in the Alps between Austria and Italy. His mummified corpse was found by hikers in 1991, lying partially exposed in the ice at the edge of a glacier. Ötzi was well equipped for his journey through the snow, being dressed in a grass rain cloak, leather leggings and a fur cap and jacket. But he also had three broken ribs and no food, which has led to speculation that he left his home in a hurry, was attacked, and then was overcome by cold.

The normal core temperature of the body, that within the deep tissues of the chest and abdomen, is 36–38 °C. Hypothermia is defined clinically as a core temperature of less than 35 °C. Its symptoms change as the body temperature falls.

Mild hypothermia is characterized by shivering, numb hands and reduced manual dexterity. Complex skills such as skiing become more difficult and you feel tired, chilled, argumentative and disinclined to cooperate with others. Mild hypothermia can be difficult to detect and is often vigorously denied by the victim. But it can be dangerous. Being unable to zip up your jacket or put on your gloves can result in further chilling or frostbite. Even a fall of as little as one degree in your core temperature slows your reaction times and can impair your judgement – indeed, mild hypothermia may be a contributing factor in traffic accidents. Motorcyclists who get chilled on long winter journeys, and

The most famous case of hypothermia: Ötzi, the prehistoric man discovered frozen deep in the ice after more than 5000 years. He was found with a wealth of possessions, including a copper axe, an unfinished bow and boots stuffed with grass for insulation. Hairs on his tools indicated he had killed red deer, chamois and ibex. And, amazingly, fungi sprouted from the grass when it was cultured.

market traders who stand around all day in the cold and then drive home, are particularly susceptible.

Moderate hypothermia occurs when the core temperature falls below 35°C. It is associated with violent shivering. As well as fine motor skills, general muscle coordination deteriorates so that you walk slowly and laboriously, stumble frequently and may fall over. Mental abilities are also affected. Speech is slurred, thinking is sluggish and the ability to make rational decisions is impaired; you may feel like lying down in the snow and going to sleep, decide to dump your rucksack because it is too heavy, or even start to remove your clothing because you are unaware of the cold. Mountaineers may fail to buckle their safety harness correctly, with tragic results. Victims become apathetic, lethargic, uncooperative, withdrawn and respond inappropriately when questioned. Often they do not remember things that happened recently.

Once the core temperature drops below 32°C, shivering stops because energy reserves are now depleted. The temperature then drops even more rapidly because of the lack of heat generation by muscle. Eventually, the person is unable to walk and curls up on the ground in a semi-conscious state, unaware of others. Consciousness is usually lost somewhere around 30°C. As one victim later related: 'I felt myself growing colder and colder. My face was freezing. My hands were freezing. I felt myself growing numb and then it got really hard to stay focused and I just sort of slid off into oblivion.'

In profound hypothermia, the heart rate slows right down, the pulse may be almost unmeasurable, and breathing becomes so shallow and erratic that it is hard to detect. The victim may breathe only once or twice a minute and their heart may beat equally slowly. Their skin is pale and feels ice-cold to the touch, their limbs are stiff and rigid and their pupils dilate and do not react to light. The person now looks dead, although they may in fact still be alive. This state is sometimes referred to as the 'metabolic icebox', because metabolism has slowed down so much that the person is almost in a state of suspended animation.

Cold slows the pulse rate because it depresses the pacemaker activity of the heart. At core temperatures below about 28°C, cardiac arrhythmias may also occur, the most serious being ventricular fibrillation – an uncoordinated twitching of the heart muscle that prevents its normal pumping action and results in death. Even if fibrillation is avoided, the heart usually stops by the time the person has chilled to 20°C.

In Arctic Waters

On 13 January 1982, Air Florida flight 90 took off from National Airport in Washington on a routine flight. Twenty-eight seconds later, the plane crashed into the 14th Street Bridge across the Potomac river. Seventy-eight people died. Not all of the casualties were severely injured in the crash; many died from hypothermia resulting from immersion in the bitterly cold waters of the Potomac. Snow, freezing conditions and darkness hampered rescue attempts, so some people were in the water for a considerable time. It was a story of tragedy and of heroism. Some victims who insisted that others be rescued before themselves were never found when the helicopter returned to pick them up.

Every year, many thousands of people die a cold-water death. Hypothermia, rather than drowning, is probably responsible for many of these deaths. Heat is lost far more rapidly from the body when immersed in water, because water is such a good conductor of heat (it has a thermal conductivity twenty-five times that of air), and immersion in water of less than 20°C leads to heat loss and eventual death from hypothermia. The colder the water, the quicker you die. In Britain, the average sea temperature in July is 15°C, and a naked man will be incapacitated within a few hours. In January, the temperature falls to 5°C and the time is shortened to thirty minutes. Immersion in water around freezing point results in hypothermia within fifteen minutes and death within thirty to ninety minutes. Without a life jacket, or in rough seas, a person usually dies even more rapidly because they will drown as soon as they lose consciousness. The British navy show their new recruits a video of the Olympic swimmer Sharon Davies swimming in freezing water to impress upon them how quickly even a world-class athlete is overcome by cold (needless to say, her attractiveness makes an equally marked impression).

The fact that cold, as distinct from drowning, is a cause of death has been appreciated by perceptive observers since antiquity, although it was not always officially recognized. Survivors of the *Titanic* disaster, for example, suggested that many casualties, who were wearing lifebelts and in calm (but icy) water, died of the cold, whereas the official inquiry blamed drowning. Scientific studies into the cause of cold-water deaths were initiated by both the British and German navies during the Second World War because of the vast number of mariners who died after escaping from their sinking ships. The most meticulous (and repugnant) of these experiments were conducted by the Nazis on the inmates of the Dachau concentration camp. The data they obtained on the thermal limits for human survival are still quoted today, and they enable experiments with volunteers (testing survival suits, for example) to be conducted within safe limits. But this raises an important ethical conundrum – is it justifiable to use the Dachau data, even when it helps understand and prevent cold-water deaths, given the horrific circumstances of the experiments and the fact that people were murdered to obtain the data?

There are many tales of individuals who have survived for a considerable time in cold water, often far longer than expected. Some are clearly statistical outliers, unusually, and providentially, insensitive to cold. In other cases, it is easier to understand how they escaped death. Tony Bullimore was competing in a round-the-world yacht race in 1997 when his boat capsized in a remote part of the Southern Ocean and he was trapped beneath it. It took four days before rescue arrived, and with the water temperature just above freezing most people held out little hope for his survival. Yet when Royal Australian Navy divers hammered on the upturned hull of his boat, Tony swam out to greet them. He was somewhat the worse for wear, but the combination of a waterproof survival suit, the hull for shelter and an insulating layer of subcutaneous fat, saved his life.

Another near victim of cold water was the eminent philosopher Bertrand Russell. In 1948, he was invited to visit Norway to give a series of lectures organized by the British Council. On 2 October he flew from Oslo to Trondheim in a flying-boat. The weather was stormy with strong winds and as the plane landed in a heavy swell, a gust of wind caught it, and blew it over on its side, so that the water flooded in through the door. Many of the passengers were unable to escape before the plane sank. Russell himself simply commented that the water was very cold. He was lucky to be rescued quickly, for the survival time of unprotected individuals immersed in the frigid waters of the North Sea is measured in minutes.

Heat loss in water is enhanced if you move around. Unless the coast is very close (less than five minutes' swim away), your best chance of survival if you are shipwrecked in cold water is to float quietly in your life-jacket until rescued; struggling or making swimming movements only serves to speed the rate of heat loss and reduce your survival time. This is because movement dissipates the thin layer of water that is warmed by the body and replaces it with a fresh layer of cold water, thus increasing conductive losses. The problem is compounded by the fact that exercise increases the circulation in the extremities, where heat loss is greatest. If you have to abandon ship, and there is time, put on plenty of thick clothes and wear gloves and footwear as the additional insulation helps prevent cold injury. Better still, wear a wetsuit or survival suit for normal clothes are not very good insulators in water.

These simple precautions may save your life. Unfortunately, they do not appear to be widely known. In 1963, the *Lakonika* caught fire off the coast of Madeira and had to be abandoned. Many of the passengers and crew ended up in the sea, where they swam around in the belief that it would help keep them warm and took off their outer clothing for

A cross-channel swimmer being covered in oil before entering the water. Like all successful long-distance swimmers, he is fairly well built which helps him withstand the cold water.

fear it would hamper their movements. For many, it was a fatal mistake: 113 people died of hypothermia.

Subcutaneous fat is an excellent insulator, so fat people tend to survive longer in cold water. Not surprisingly, most successful swimmers of the English Channel are fairly well built. Usually, swimmers attempt the 34.5-kilometre (21.5-mile) crossing in August or September when the water temperature is warmest, albeit still a chilly 15 to 18 °C. They take between nine and twenty-seven hours to swim across, which is significantly longer than the calculated survival time for a human immersed in water at that temperature. Several things contribute to their success – the exercise generates a significant amount of body heat, they usually have a substantial layer of subcutaneous fat, and they are fed at regular intervals (they don't stop swimming while eating, in case they get cramp). All the same, many have to give up because of fatigue or hypothermia,[2] and in August 1999, an experienced long-distance swimmer died while attempting the crossing.

Ice-diving is one of the latest 'extreme sports', an ultimate adventure that cannot fail to raise the adrenaline level of even the most jaded individual. After dynamiting a hole in a frozen lake, enthusiasts dive into the freezing water and swim below the ice. A fit person wearing a wetsuit can survive underwater at around 1 °C for about twenty minutes (with the assistance of an air supply) before they become dangerously cold. The endurance record for an unassisted (breathheld) ice-dive is held by Fabrice Bougand, a Frenchman who spent two minutes and thirty-three seconds under the ice, at a temperature of 10 °C.

Ice-divers should be wary, however, for cold water can kill instantly. There is, in fact, a very real foundation to the common excuse that diving into cold water 'will kill me'. A number of cases have been reported of fit young people who have floated up to the surface dead (or sunk to the bottom) within a minute of two of plunging into an icy lake. Even good swimmers have been affected. Why this should happen is still not clearly understood, but several physiological responses may be involved. Shock and pain slow the heart, and may precipitate a fatal arrhythmia, while a reflex gasp for breath triggered by the cold can be fatal if taken underwater. Cold shock also causes hyperventilation, washing carbon dioxide out of the blood and decreasing its acidity – the result is tetany of the muscles (which prevents coordinated swimming), loss of consciousness and rapid death from drowning.

Cold water kills in several ways. Death may occur within seconds of entering the water, after a few minutes of swimming, or much later, when the body has chilled and consciousness has been lost. It may also occur following rescue. Accounts of the sinking of the German battleship *Gneisenau* during the First World War refer to most survivors dying on board the rescue ship despite having been fished out of the water alive and apparently unharmed. There are also many contemporary reports of shipwrecked sailors and fishermen being conscious when taken out of the water, but losing consciousness and dying soon after rescue. The cause of this post-rescue collapse is still controversial, but it seems to be due to a combination of cold and changes in hydrostatic pressure on removal from the water, as described in Chapter Two.

Losing the Balance

Hypothermia occurs whenever heat loss exceeds heat gain: it need not necessarily result from wintry conditions. Elderly people with inadequate food or heating are particularly susceptible, especially if they suffer injuries that immobilize them. Their body temperature falls steadily over one or two days, producing a gradually increasing state of confusion, incoordination and stupor. Malnourished patients, particularly if they are children, have to be kept in environments that are often uncomfortably warm for their carers, because their low metabolic rate predisposes them to hypothermia. Drugs that increase heat loss can produce hypothermia even at relatively warm environmental temperatures.

Hypothermia can also result from a combination of exercise, inadequate food and alcohol. Exercise depletes the body's store of carbohydrate, so that blood sugar concentration tends to fall. Drinking alcohol exacerbates the problem by lowering the blood sugar still further, because glucose is needed to metabolize the alcohol. Hypoglycaemia (low blood sugar) greatly reduces the body's response to cold and because blood flow to the skin is not shut down, heat is lost at an alarming rate. In these circumstances, the core temperature can fall rapidly even when it is not particularly cold: for example, exposure to an air temperature of 20 °C has been found to chill the body to 33 °C within eighty minutes. A vigorous walk for a couple of hours on an empty stomach, followed by several slugs of whisky to 'help warm you up', can be a dangerous combination.

Life After 'Death'

'Nobody is dead until they are warm and dead,' is the doctor's maxim. Almost every year there are tales of the 'miraculous resurrection' of victims of cold, because people suffering from deep hypothermia may appear dead despite being still alive. In February 1999, avalanches swept through the Swiss and Austrian Alps. Among the many casualties was a four-year-old child who lay buried in snow for two hours. Pronounced clinically dead when he was dug out, the rescue teams were able to resuscitate him and within a couple of days he was once again playing unconcernedly with his toys.

The lowest core temperature recorded for an individual who survived accidental hypothermia is 13.7 °C. This was a 29-year-old Norwegian woman who fell over while skiing down a waterfall gully, became wedged between the rocks and thick ice and was continuously drenched by a flow of icy water. Her companions were unable to extricate her and by the time the rescue team arrived, an hour and ten minuites later, she was clinically dead. Nevertheless, she was immediately given cardiopulmonary resuscitation and transferred to Tromsø University Hospital where an experienced resuscitation team were able to revive her. Five months later she had almost completely recovered.

Small children have also been revived after being totally submerged in icy water for many minutes without breathing, because the cold slows their metabolic rate so much that very little oxygen is required. A typical case is that of a five-year-old boy who fell through the ice when walking on a partially frozen river and was trapped beneath it for forty minutes before being rescued by frogmen. There were no airpockets between the ice and the water and he appeared to have been submerged underwater for the whole time. When he was fished out he had no pulse, was not breathing, was blue-grey with cold and had a core temperature of 24 °C. After two days on a respirator in hospital, he recovered consciousness and started to talk. Eight days after his accident he was allowed home. He was very lucky, for he made a full recovery, and showed no sign of brain damage. Not everyone will survive such a ordeal. Children usually do best, because they are so small that they are quickly chilled, their oxygen demands fall rapidly, and they enter a state of suspended animation.

The quickest method of rewarming someone suffering from moderate hypothermia is to immerse them in a warm bath. Victims of deep hypothermia are warmed by giving them warm air to breathe, puffing warm air on their skin, and by removing their blood via a vein and circulating it through a heat-exchanger before returning it to the body. Great care must be taken when rewarming severely cooled individuals, as cardiac arrythmias may occur. This is a particular problem when rewarming those whose heart has stopped.

On Chapped Hands and Cold Feet

As a small child I was told not to warm my cold hands on the school radiators for I would get chilblains. Little heard of today, chilblains seemed to be a common complaint in my youth, although fortunately I never got them (despite not obeying instructions). Chilblains are red and itchy patches of skin that are found most commonly on the fingers, toes, cheeks and ears. They are caused by repeated exposure of bare skin to temperatures below 15°C, which leads to permanent damage to the fine capillaries. Women and children are particularly susceptible. Chilblains are more frequent in humid conditions like those found in Britain than in cold, dry climates. The reduction in their incidence in Britain today is probably the result of increased affluence, which has resulted in adequate clothing and the widespread introduction of central heating.

Raynaud's syndrome is a condition in which the fingers (or toes) turn white, then blue and then red on exposure to cold. This is because the blood vessels initially constrict so strongly that almost all blood flow ceases, and then slowly dilate again. It can be quite painful when the blood reaches the bloodless fingers once more. Strangely, Raynaud's syndrome (like chilblains) is more common in countries with relatively mild winters, like Britain and Italy, than in Canada and Sweden, perhaps because the more severe climate ensures that people take better precautions. In Britain, for example, schoolchildren play outside in winter and are thus chronically exposed to cold.

Trench foot was a dreaded corollary of the First World War. Over 29,000 cases were documented in the British army in 1915. It is caused by prolonged exposure to cool wet conditions, and the conditions in the trenches were ideal. Rain filled the trenches, so that the men were constantly wading in water and thick sticky mud, and freezing winds stiffened their wet clothing into icy boards.

Trench foot continues to remain a problem. During the Falklands campaign of 1982, it accounted for 14 per cent of British casualties and in 1988, 11 per cent of one US Marine unit suffered from the condition. Sea kayakers whose hands and feet may be exposed to water for long periods are particularly susceptible, mountaineers may succumb if their socks become soaked by excessive perspiration or powder snow gets into their boots and melts, and it can also happen to people whose occupations require them to stand about with wet feet in wintry conditions. Each year, scores of people attending the Glastonbury festival who arrive with inadequate footwear develop trench foot, for even in June the English weather can be cold and wet, and the ground becomes a quagmire.

Trench foot is a local, non-freezing cold injury that results from prolonged exposure to cool, wet conditions. In fact, the temperature does not have to be very cold – standing around in water at 10 °C for twelve hours or more is sufficient. Wet feet lose heat very quickly – about twenty-five times more rapidly than if they were dry – so that the blood vessels supplying the feet constrict to reduce heat loss. When the circulation is reduced in this way, the tissue begins to die for it no longer receives the oxygen and nutrients it needs, and toxic metabolites accumulate. Trench foot is particularly insidious because the deeper tissues, such as muscles and nerves, may be affected long before there is any noticeable skin damage. The affected limb is cold, appears a pale mottled colour and feels numb. On rewarming, the skin becomes purplish-red, swollen and extremely painful. Some victims say it is like 'electric shocks running up the legs from the toes'. Blisters, ulcers and gangrene may develop and in very severe cases the entire foot may die and have to be amputated.

Simple but effective measures prevent trench foot. The key is to keep the feet dry at all times and to avoid anything that restricts the circulation, such as remaining immobile for long periods in a cramped position. Unfortunately, as may be imagined, it is not always possible to achieve either of these in military operations.

Frostbite

If the skin cools to temperatures around zero, frostbite can occur due to freezing of the tissues. It occurs most commonly in the extremities –

A Sherpa with swollen and blistered hands caused by frostbite

such as the ears, nose, fingers and toes (and, spoiling the mnemonic, also the cheeks). In mild cases, only the outer layers of the skin freeze. Frostnip, as this is sometimes called, is characterized by a white, waxy-looking skin and a loss of sensation. It is similar to sunburn and other first-degree burns, and, on rewarming, the frozen skin turns bright red and is later sloughed off. Superficial frostbite is more serious for, in addition to the skin, the underlying tissues freeze. On rewarming, the skin turns blue-purple and swells. Within a day or two, blisters may form and a hard black carapace appears. If the individual is lucky, the skin heals beneath this layer, which is finally sloughed off to reveal the new skin. But it can be very painful, as Apsley Cherry-Garrard describes:

> 'The temperature was -47°F and I was fool enough to take my hands out of my mitts to haul on the ropes to bring the sledges up. I started away with all ten fingers frostbitten. They did not really come back until we were in the tent for our night meal, and within a few hours there were two or three large blisters, up to an inch long, on all of them. For many days, those blisters hurt frightfully.'

The most serious form of frostbite results from freezing of the deeper tissues, such as the muscles, bone and tendons. Deep frostbite almost invariably results in permanent tissue damage and may require eventual amputation. Many polar explorers and mountaineers have lost their fingers or toes to frostbite. Horrific frostbite injuries were sustained by Beck Weathers, a member of one of the ill-fated Everest expeditions in May 1996 that encountered a ferocious storm. Left for dead on the ledge of the Kanshung Face in comatose state, without his right glove, his face caked with ice, 'as close to death as a person can be and still be breathing', Beck miraculously refused to succumb.[3] After twelve hours of semi-consciousness, he slowly became aware of his predicament and realized that 'I was in deep shit and that the cavalry wasn't coming so I had better do something about it myself'. He staggered back to camp, critically ill. He would ultimately lose his right arm below the elbow, all the fingers and the thumb of his left hand, and his nose. But he kept his life and, as his friends testify, his sense of humour.

As tissue freezes, ice crystals form in the cells and fluids that bathe them. If freezing is slow, ice crystals first appear in the extracellular fluids. This increases the concentration of the solution that remains unfrozen, and drags water out of the cell by osmosis (the tendency of water to move from a solution of high concentration to one of low concentration). Consequently, the cell shrinks and the concentration of the salt solution inside it rises. Because proteins are permanently damaged by high salt levels, this results in cell death. When freezing is rapid, needles of ice may crystallize inside cells, puncturing the cell membrane. If ice crystals rub together, they may physically tear the cells apart, which is one reason it is not advisable to rub frostbitten areas.

Further injury occurs on rewarming. The cells lining the walls of the finest blood vessels are particularly susceptible to damage and on rewarming they become leaky. Fluid seeps out of them, causing swelling of the surrounding tissue. Sludging of the red blood cells left

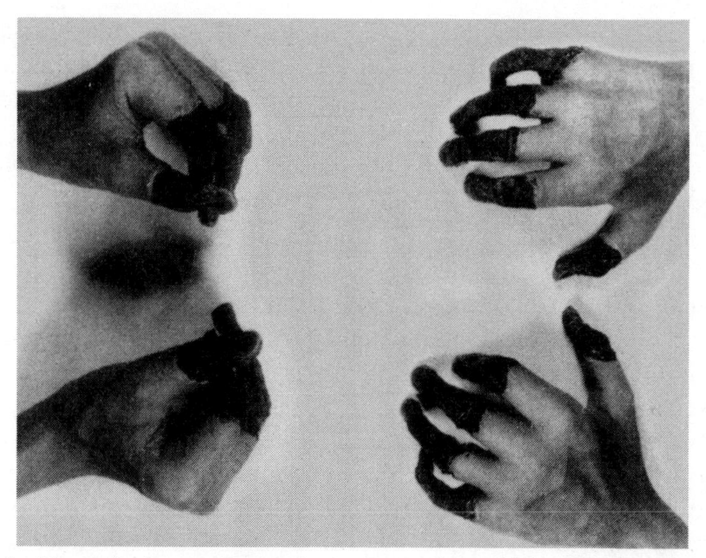

Deep frostbite of the fingers, showing that movement is possible because the muscles are unaffected and the tendons are still intact.

behind in the capillaries reduces the blood flow, which in turn decreases the supply of oxygen and nutrients to the tissues downstream and ultimately leads to their death. The extensive damage that can occur on rewarming means it is wise to keep severely frostbitten tissue frozen until it can receive medical attention. Thawing and refreezing can be catastrophic.

Eskimos and Explorers

It is well established that some people are comfortable in conditions that most of us would find unacceptably cold. As Darwin recounts, the Yaga Indians of Tierro del Fuego lived through the snow and ice of the Patagonian winter without any clothing (they had fires, however, which gave the land its name). The Australian Aborigines and the Kalahari Bushmen dwell in desert areas in which the temperature falls precipitously at night and may drop below freezing in winter. Despite the cold, aborigines traditionally slept naked on the ground with only a windbreak for shelter. Physiological studies have shown that they allow their core temperature to cool at night to around 35°C and their skin temperature also falls. The Kalahari Bushmen show a similar response. In contrast, white Europeans exposed to the same conditions maintain their temperature at 36°C by shivering and thrashing about continuously, and thus are unable to sleep. Yet even among Europeans, individuals differ in their ability to cope with cold. In my view, my sister's home is freezing, while she finds mine uncomfortably hot.

'Birdie' (H.G.) Bowers, a member of Scott's ill-fated last expedition (in 1911), was noted for his extreme hardiness. On the winter journey to Cape Crozier to collect the eggs of the Emperor penguin, Bowers slept soundly in temperatures below -20°C without the eiderdown lining of his fur sleeping bag, whereas his companion, Apsley Cherry-Garrard, suffered 'a succession of shivering fits which I was quite unable to stop and which took possession of my body until I thought my back would break, such was the strain placed upon it'. Unlike Cherry-Garrard, Bowers was also never bothered by frostbitten feet. Scott commented that he had 'never seen anyone so unaffected by cold'.

Why was Bowers so insensitive to cold? One possible explanation is that every morning, to the fascinated horror of his companions, he would strip naked in the freezing Antarctic air and douse himself with buckets of icy water and slush. Several studies have shown that intermittent exposure to cold seems to cause some degree of cold adaptation in humans. Daily immersion of nude volunteers in water of 15°C for thirty to sixty minutes over several weeks, for example, resulted in greater tolerance and less discomfort when they were subsequently exposed to Arctic conditions. One of the survivors of the great retreat from Moscow in 1812, Lieutenant J.L. Henckens related that he also 'managed to keep warm by rubbing myself with a great deal of snow, a commodity to be found in profusion'.

All this suggests that Bowers's regular icy ablutions may have been responsible for his extreme tolerance of cold. It may also account for the proverbial hardiness of the Spartans, and of British public schoolboys, who reputedly bathed daily in cold water. A similar physiological adaptation probably underlies the ability of some people to work for long hours with their hands in water so cold that others cannot endure it. Fishermen, Eskimos and American Indians, for example, can continue to maintain a continuous circulation to their extremities even in the cold, as described earlier. These findings have led some authorities to suggest that a regime of regular icy baths may assist in preadaptation to cold environments. Others, however, have argued that any possible advantage is more than offset by the deleterious effect on morale. In general, the attitude is that of one US airman who, while advocating that infantry should be trained to operate in light clothing to enhance their mobility, baulked at the idea when it was suggested for himself.

Cold stimulates the appetite, and the increased food intake leads to a higher metabolic rate and enhanced heat production. Eskimos have a basal metabolic rate that is up to 33 per cent higher than that of Europeans, mainly because of their traditionally high protein diet, which includes as much as a pound of meat a day. This, in part, accounts for their greater tolerance of cold. Chronic cold may also increase subcutaneous fat. Seasonal fluctuations in weight have been noticed in Britain, with an increase in winter and a decrease in summer, and it is said that the thighs of girls in temperate countries became fatter during the miniskirt era (although a sceptic might argue it was simply that their size was more apparent). In any case, these changes in body fat are too small to have any significant effect on thermal balance and there is no evidence that populations living in cold environments are fatter than those in tropical ones. Races that have evolved in different climates may have different shapes, however, as discussed in Chapter Three.

The Benefits of Cold

Cold is not always detrimental. During the Falklands conflict, it was noticed that many casualties inexplicably survived severe injuries, such as the loss of a limb, despite being unable to reach a field hospital for many hours. Subsequent studies suggest that the intense cold markedly reduced blood loss from their wounds (as in the case of Thirion's horses) and caused a mild hypothermia that decreased their bodies' demand for oxygen and enabled them to survive even with a reduced blood volume.

Low temperatures are sometimes used deliberately during operations to slow the body's metabolic rate and so reduce the oxygen requirement of the tissues. This enables the blood flow to be interrupted without damaging the tissue. In cardiac surgery, for example, the heart may be stopped for up to an hour by the administration of cold solutions at around 4 °C (the rest of the body is perfused with warm blood by a heart–lung machine). For certain neurosurgical operations, the brain is cooled, enabling the local circulation to be stopped for up to fifteen minutes. Keeping a cool head may also help prevent babies who are starved of oxygen during a difficult birth from later developing irreversible brain damage. Most damage actually occurs one or two days after birth and in animals it can be prevented by chilling the brain postnatally. Experiments are currently being carried out on human babies to determine if postnatal fitting of a water-chilled helmet, which cools the baby's head by about 3 °C, reduces brain damage.

On Penguins and Polar Bears

Humans evolved on the plains of Africa and our capacity to cope with cold is limited. By contrast, many animals are perfectly adapted to cold environments. They are well insulated by thick fur or subcutaneous fat; they tend to be large and have short extremities, which helps reduce their surface-area-to-volume ratio, and thus their heat loss; and many have antifreeze proteins in their blood and tissues. Others opt out of

The Arctic fox survives happily at -50°C. Its thick winter coat provides effective insulation and it sleeps curled into a tight ball with its nose and feet on the inside. Its compact body, short legs and small ears also help to conserve vital heat.

active life, allowing their body temperature to cool (or even freeze) and their metabolic rate to drop, hibernating until the inclement weather passes. So effective are these strategies that for many animals the real problem of life in a cold climate is not the temperature but the limited food supply.

Fur (or feathers) keeps an animal warm because air is trapped within the hairs and provides an extra layer of insulation. Fluffing up the coat by erection of the hairs or feathers increases the amount of the trapped air and reduces heat loss. Air is a very effective insulator. It accounts for the fact that many layers of clothing are warmer than one thick layer and that a string vest, which is mostly holes, keeps you warm – the air pockets between the strings are responsible. (The string vest was developed for the British Graham Land expedition to the Antarctic in 1920–22 and was still around in my youth but it is no longer in common use.)

The protective effect of animal fur drops when the wind speed increases because the trapped layers of warm air are disturbed. Logically, it would be far warmer to wear a fur coat inside-out, as the Eskimos do, but this option is not available to the animal itself. It is clear that in the developed world most fur coats are worn as a fashion item or a status symbol, rather than for practical purposes, for only sheepskin coats are worn inside-out.

Fur and feathers are very good in air, but useless in water. The air escapes from between the hairs and its insulating qualities are lost. In water, fat or blubber is a far more effective insulator. Seals have substantial layers of fat under their skins; as do polar bears, who also spend a considerable amount of time in icy seas. And fat humans may be expected to survive longer than thin ones when immersed in cold water.

A man standing barefoot on an ice floe would rapidly get frostbite, yet penguins do so all their lives without injury. Penguins' feet don't get frostbite because they never cool to the temperature of the ice; blood flow to the feet is adjusted to keep them a few degrees above zero and prevent them freezing. When the air temperature drops below -10°C, Emperor penguins reduce their points of contact with the ground by standing on their heels and tails, holding their toes up and their flippers close to the body. At first glance, it might seem strange that penguins do not have better insulated feet; however, penguins are exceptionally well insulated overall, which makes it difficult for them to get rid of the excess heat generated when they exercise. The feet are one of the few regions of the body where this is possible.

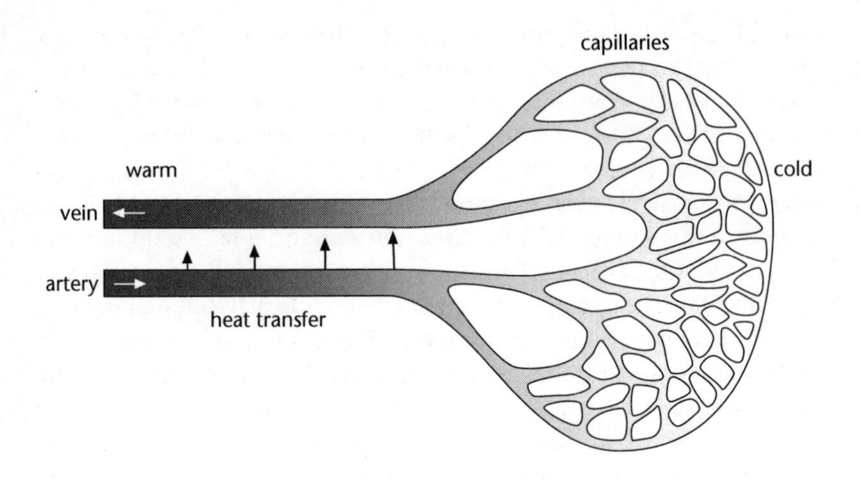

When a peripheral artery runs alongside a vein there will be a net thermal gradient between them because the blood leaving the body's core will be hotter than that returning from the cold skin. This will result in the transfer of heat from artery to vein, and produce a thermal short-circuit that retains the heat within the body's core and reduces peripheral heat loss. Even in humans, such heat exchange occurs. The arteries run deep in the tissues, while two sets of venous vessels are present – one that runs adjacent to the arteries and one that lies just under the surface of the skin. Shunting the blood from the peripheral vessels to the deeper ones helps conserve heat. In some animals, this simple counter-current arrangement is elaborated into a vascular heat exchanger (the *rete mirabile*), consisting of hundreds of intermingled small arteries and veins (see above). The *rete mirabile* of the bluefin tuna was first observed by the French naturalist Georges Cuvier in 1831.

It is not easy for a fish to maintain its body temperature higher than that of the surrounding water because the rapid and extensive blood flow across the gills needed for respiration inevitably leads to excessive heat loss. Tuna fish and sharks, however, have evolved a vascular countercurrent heat exchanger that enables them to maintain their muscles up to 20°C hotter than that of the rest of the body. Like the *rete mirabile* of the antelope described in Chapter Three, this consists of a network of hundreds of intertangled small arteries and veins, but in this case heat is transferred from the warm blood leaving the working muscle to heat the cooler incoming blood. The warmer muscle accounts for the ability of tuna fish to swim at speeds as high as 18 kilometres an hour. Similar countercurrent heat exchangers are found in the flippers of seals and dolphins, and the tail flukes of whales, where they help to prevent heat loss into the ice-cold sea water. Wading birds that stride through the water all day on long, thinly insulated stilt-like legs also possess *rete mirabile* in their legs. This helps explain why they suffer no ill effects, while a man with his feet immersed in icy water would get trench foot.

Animals also adapt to chronic cold by biochemical changes within their cells that enable them to operate at lower temperatures. Human nerves and muscles cease to work when they are cooled below 8 °C, but those of Arctic animals continue to function at temperatures close to zero. The reason for this difference lies in the nature of the fats (technically, lipids) found in their cell membranes. Most types of animal fat become hard and brittle when cooled, but those found in the legs of seagulls and animals that live in cold climates have a melting point that varies with the distance from the body core. Fat extracted from caribous' feet remains fluid at cold temperatures and is used as a lubricant by Eskimos, while that from the upper leg is solid even at room temperature and is used for food. Closer to home, neat's-foot oil, which comes from the feet of cattle, can be used to keep leather supple in the cold. Changes in the amount of saturated fat in the cell membranes account for these physical differences. Saturated fats, like butter, are hard at low temperatures, while unsaturated fats, like olive oil, remain soft or liquid. Amazingly, the membrane of the same nerve cell can change its lipid composition throughout its length, having fewer saturated fats in the extremities and more when it lies within the bulk of the animal's body. This ensures that the fluidity of the membrane is constant throughout the length of the cell and maintains nerve and muscle function even in the cold.

Like humans, animals modify their behaviour to cope with cold. The Emperor penguin *(Aptenodytes forsteri)* lives in the Antarctic and is exposed to some of the most extreme conditions on Earth. It breeds in the depths of winter in air temperatures as low as -30 °C, which are further reduced by freezing winds that can reach speeds in excess of 200 kilometres an hour. The communal rookeries are not sited on the floating ice but on the permanent ice shelf, many kilometres from the open sea. There is no food in these frozen wastes so the penguins suffer an enforced fast during the breeding period. In March, when the ice collar around Antarctica is at its narrowest, both male and female birds begin the long march to the rookery. After laying a single egg in late May or June, the female returns to the sea to feed, leaving her mate incubating the egg in his brood pouch until she returns around two months later. The male endures the worst conditions of the Antarctic winter. During this time he does not feed and must survive entirely on his reserves of body fat. When the female returns to relieve him of his duty, he may have lost as much as 40 per cent of his body weight. His long fast is not over, however, for he must walk to the open sea before he can feed, which now may be as much as 200 kilometres away because of fresh ice formed during the winter. The male must therefore fast from the moment he leaves the sea until he returns, a period that can exceed 115 days.

Scientists have calculated that the heat that can be produced by his stored fat is not sufficient to maintain the penguin's body temperature at its normal level of 38 °C in the intense cold of the Antarctic winter. So how do the Emperor penguins survive? The secret lies in their social behaviour. The adults, and also later their chicks, huddle close together in large groups of several thousand individuals. This reduces the surface area exposed to the freezing air and so conserves heat. These giant rafts

Emperor penguin chicks huddling together for warmth. The Emperor penguin is famed for its size and for its ability to withstand extreme cold. The most gripping of all polar adventures is that recounted by Apsley Cherry-Garrard in *The Worst Journey in the World*, his classic account of the search for the egg of the Emperor penguin conducted during the depths of the Antarctic winter, in continuous darkness and at temperatures below -70°C.

of birds are continuously moving as the penguins on the outside slowly shuffle their way to the centre, displacing the warmer birds found there to the outside.

Huddling is not confined to penguins. Bees also cluster together at low temperatures, a behaviour that enables the group to overwinter in temperatures that would cause the death of a solitary individual. As the temperature falls they pack closer together, thereby reducing heat loss even further. Within the centre of a cluster, it can be as warm as 30 °C despite an ambient air temperature as low as 2 °C. The outer parts of the cluster cool to around 9 °C, just above the temperature at which the bees become comatose. As with penguins, there is a constant circulation of cooler bees from the periphery to the warm heart of the cluster. A group of humans caught out in the cold would do well to emulate this behaviour. Indeed, the old practice of 'bundling', still seen in some pre-industrial populations, in which several family members sleep in the same bed, serves the same purpose, although it is less efficient because rotation of the central group members does not take place.

Insects can only fly when their muscles are warm enough; cold flight muscles simply don't work. The Wakamba tribe of Kenya are said to take advantage of this fact by robbing wild hives at night when the cold partially incapacitates the bees. Because their body temperature is close to ambient at rest, insects must warm up their muscles before their first flight of the morning. Many do this by simply basking in the sunshine, but others, like moths and bees, generate heat internally by rapidly contracting their flight muscles. Moths vibrate their wings silently but bees warm up their muscles by contracting them without visible movements. Bumblebees also have a 'fur coat' on their thorax, which cuts heat loss in half. In contrast to moths, most butterflies are grounded without the sun's heat and are only seen dancing over the flowers on warm sunny days. In the early morning, they tilt their wings towards the sun; these act like solar panels, collecting the heat of the sun's rays and transmitting it to the flight muscles. Only then can the butterfly take off. When the sun goes behind the clouds, the temperature drops by one or two degrees and the butterflies are grounded once again.

Like insects, lizards are heliotherms, using the sun to heat themselves directly. When cold, they orientate themselves at right angles to the sun's rays to absorb maximum heat. In the desert, where the ground is warmer than the air, they snuggle against the ground to absorb its warmth, and on cold rocky mountain slopes they use dead grass as insulation. When it gets too hot, they take avoiding action, by retreating to the shade or underground. Large animals take much longer to heat up, which probably explains why all the large reptiles – the crocodiles, monitor lizards, Komodo dragon, and giant tortoises – live in the tropics. Some lizards have specialized pigment cells in the skin that help to regulate heat gain from the environment. In the cold, the black pigment cells expand and increase the rate at which the animal absorbs heat, while in hot sun they contract, exposing adjacent cells that reflect infrared rays. A sluggish lizard is vulnerable to fast-moving predators, and the earless lizard has developed a remarkable adaptation to decrease its risk of danger. It thrusts its head out of the burrow in the morning, exposing a large blood sinus in its head. Once this has absorbed enough heat to raise its body temperature, the lizard emerges, able to take off at top speed if necessary.

Like humans and insects, snakes generate heat by muscle contraction. In 1832, the French scientist P. Lamarre-Picquot suggested that the Indian python coils around its eggs and warms them with its own body heat. His idea received little credence at the time and was rejected by the French Academy of Sciences as 'hazardous and questionable'. Nevertheless, Lamarre-Picquot was correct. Studies in the 1960s showed that, by contracting its muscles, the python is able to keep its body temperature about 5 °C higher than that of the ambient air temperature.

The most extreme examples of behavioural adaptation to cold are migration and hibernation. Small mammals are unable to maintain a core temperature of 37 °C in very cold environments, as they simply cannot eat enough to obtain the fuel they require. Instead, they opt out of homeothermy and hibernate until the climate is more clement. Because cold tissues require less energy, they allow their metabolic rate to fall, thus conserving their energy reserves, and their body temperature concomitantly declines from 37 °C to the ambient level. Heart rate, respiratory rate and biochemical reactions in the tissues also decrease. Hibernation is a highly regulated process – it is a resetting of the thermostat to a much lower level, rather than a failure to thermoregulate. If the ambient temperature falls below 2 °C, the animals actively generate heat, keeping their temperature between 2 and 5 °C to ensure that they do not freeze. In very cold weather, they may even wake up. Hibernation is initiated by the changes in temperature, day length and availability of food that signal the onset of winter. In spring, arousal occurs rapidly and the core temperature can rise by as much as 30 °C in ninety minutes. Rapid awakening is achieved by hormones that activate brown fat metabolism, which warms the animal up.

Small passerine birds migrate to warmer latitudes in winter, or descend from the mountains to the plains. While this helps avoid the cold and reduced food supply, it also entails physiological adaptations for the long migration. Most small birds must fatten up beforehand because of the high energy cost of sustained flight. Many must also stop *en route* to refuel, because the weight restrictions imposed by flight mean they cannot store enough fuel for the whole of their journey. Ironically, humans, who also migrate to more pleasant climates by taking winter holidays in warm sunshine, often try to *lose* weight before migration.

Life at the Poles

Life at the poles, or on the mountain tops, brings many problems in addition to cold. During the summer months, the sun never sets at the poles but simply circles round the sky each day. On clear days, the radiation can be intense and may cause severe sunburn. Reflections from snow and ice dazzle the eyes so that goggles are essential to prevent snowblindness, a kind of sunburn of the eyes, which makes them feel as if they are full of sand and makes blinking extremely painful. White-outs, when land and sky merge imperceptibly into one another, make even walking difficult. With no shadows to provide contrast, surface irregularities are undetectable, for snow and ice are the same shade of blue-white as the hole that lies beside them. Thus it is possible to stumble into a crevasse or walk into a waist-high chunk of ice without warning. The perennial problems of finding food and water become even more difficult and, without adequate support facilities, life in the cold is hazardous for humans, as many polar explorers and mountaineers have found to their cost.

5

LIFE IN THE FAST LANE

'Now! now,' cried the Queen. 'Faster! faster!'

LEWIS CARROLL, *Through the Looking Glass (and What Alice Found There)*

Sir Roger Bannister at the finish of the first four-minute mile

ONE WINDY AFTERNOON in May 1954, a young runner arrived at the Iffley Road sports grounds in Oxford to take part in a contest between Oxford University and the Amateur Athletics Association. It was not an auspicious day for a world speed record because gale-force winds had been blowing for several days. Yet that afternoon, Roger Bannister ran a mile in under four minutes. Bannister, a former Oxford medical student, and already famous as a miler, was running for the AAA team, together with his friends Chris Chataway and Chris Brasher. The two Chrises played an important part in his success by acting as pacemakers and ensuring that Roger, who 'felt tremendously full of running', did not over-exert himself early on and so be unable to sustain the pace throughout the race. Bannister crossed the finishing tape in 3 minutes and 59.4 seconds. He collapsed after his stupendous effort, writing in his autobiography that he felt 'like an exploded flashlight with no will to live. . . Blood surged from my muscles and seemed to fell me. It was as if all my limbs were caught in an ever-tightening vice.' His paralysis was only temporary. Moments later his time was announced, the crowd erupted in a roar of excitement, and Bannister and his friends ran round the track again in triumph. It was hailed as one of the greatest athletic achievements of the century. Sir Roger went on to have a distinguished career as a neurologist, but it is for his historic race that most people will remember him.

When Roger Bannister first ran a mile in under four minutes it was widely believed that it could not be done. His demonstration that this was not the case opened the minds of other athletes and within a few months his record was broken. There are now many men (but as yet no women) who have not only equalled his magnificent achievement but even surpassed it.

The current world record is 3 minutes and 43.13 seconds and was set by Hicham El Guerrouj of Morocco on 7 July 1999. Yet his time was only 1.26 seconds faster than that of the previous record-holder, Noureddine Morceli. Other world records are also constantly being broken, but by increasingly smaller amounts. The most recent records for the 100-metre footrace are 9.85, 9.84 and 9.79 seconds set by Leroy Burrell in 1994, Donovan Bailey in 1996 and Maurice Greene in 1999. This is an advance of only 0.6 seconds in five years and it raises the question of whether the current world record is close to the human speed limit. This chapter considers the physiological constraints on speed, stamina and strength and what sets the limit to how fast we can run, how far we can jump, and how heavy a weight we can lift.

A Question of Energy

As a runner waits at the starting block for the gun to go off, various anticipatory mechanisms switch on, preparing their body for the coming race. Adrenaline levels in the bloodstream rise, quickening the pulse and causing the heart to contract more forcefully. As a result, the amount of blood pumped each beat increases. Breathing becomes deeper and may speed up slightly. The muscles tense and blood is switched away from other tissues to increase the supply to the leg muscles. All these changes occur even before exercise actually starts.

Crack!! The starter gun sets the runners on their way. As they leap forward, there is an immediate jump in the rate and depth of breathing. The heart rate rises rapidly to its maximum level, and the volume of blood ejected each beat increases. Haemoglobin in the red blood cells surrenders more of the oxygen it carries to the muscles, in response to their enhanced demand. As a sprinter races forward at top speed, his muscles generate a considerable amount of heat and the skin flushes as blood is diverted to the surface of the body to help cool him down. A few seconds into the race, the immediate energy stores are depleted, lactic acid begins to accumulate in the muscles and the athlete starts to run short of oxygen. Pressed further, his body reaches the point of exhaustion because it is unable to supply fuel and oxygen to the muscles sufficiently rapidly. If he does not slow down, things begin to fail. The heart rhythm becomes less regular, cardiac output declines, the oxygen content of the blood may fall and the body temperature rises. The athlete becomes clumsy, uncoordinated and is close to collapse.

No one, not even an elite athlete, can run very far at their maximum rate. Distance running requires different skills. In order to cover many miles in the shortest possible time, the pace must be slower, for only then can the muscles be supplied with the oxygen and fuel they need without sustaining an oxygen debt. The marathon runner must balance speed against stamina.

The key to both speed and stamina is the rate at which energy – in the form of adenosine triphosphate – can be generated to power muscle contraction. Adenosine triphosphate, commonly abbreviated as ATP, is a very special molecule. It is the energy currency of the cell, the biochemical fuel that powers the cells of all living organisms be they bacteria, plants or animals. ATP consists of an adenosine head with a tail of three phosphates. The phosphate tail is the most important part of the molecule because the phosphate groups are attached by high-energy chemical bonds. Splitting off the terminal phosphate releases the energy stored in the chemical bond, which is then available for muscle contraction. Muscle contraction is not very efficient, however, and only about half of the energy stored in ATP is actually used for work. The rest is dissipated as heat, which is why you get hot when you run.

In spite of its importance, very little ATP is stored in muscle – only enough for one or two seconds of strenuous exercise. Thus ATP must constantly be replenished, by adding back a phosphate to the adenosine diphosphate (ADP) molecule that is produced when the terminal phosphate of ATP is removed. The immediate source of high-energy phosphate for ATP regeneration is creatine phosphate, which is present in muscle in fairly large amounts. Creatine phosphate is another energy-rich compound but, unlike ATP, it cannot be used directly for muscle contraction. Instead, it transfers its high-energy phosphate to ADP, thereby producing ATP. The amount of creatine phosphate in muscle is sufficient for about six to eight seconds of all-out exercise – for a 50-metre dash or a tennis serve that sends the ball hurtling across the court at a sizzling 130 miles an hour. But it, too, is soon depleted.

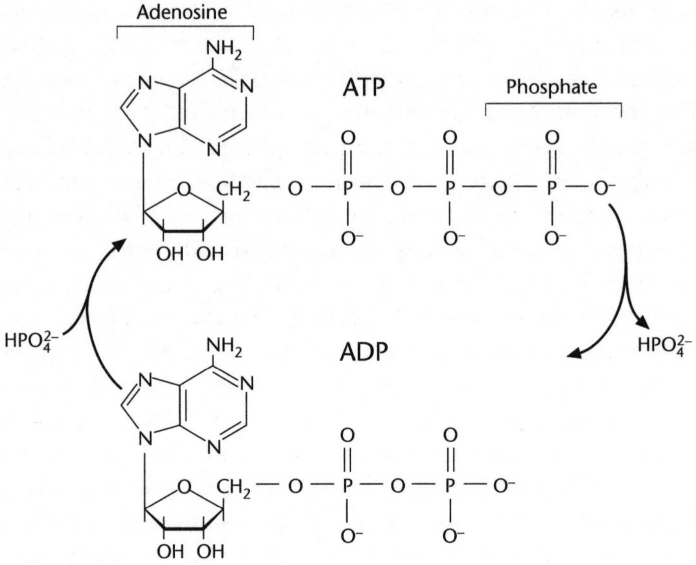

ATP – the fuel of life

Once creatine phosphate is exhausted, ATP must be replaced by the metabolism (breakdown) of carbohydrate or fat. Muscle contains a limited store of carbohydrate in the form of glycogen (animal starch), which normally makes up 1–2 per cent of the muscle mass. This lasts for about an hour of exercise, after which glucose and fat must be mobilized from storage depots in the liver and adipose tissue. Oxygen is always needed to metabolize fats, but carbohydrate can be broken down either by oxygen-requiring aerobic pathways or by anaerobic pathways that do not utilize oxygen. Because of the need for oxygen, aerobic metabolism cannot supply energy as quickly as anaerobic metabolism. This means that fats are not as immediate an energy source as glycogen or glucose. Fats also need more oxygen to break them down. For fast running, therefore, carbohydrate is the better fuel.

Anaerobic (without oxygen) metabolism of glycogen and glucose serves as a short-term means of resupplying ATP during heavy exercise and is of critical importance in sports like football, where brief bouts of high-intensity exercise that exhaust the immediate supply of ATP are common. Anaerobic metabolism cannot continue for ever, however, because it produces lactic acid, and the accumulating acid eventually impedes muscle activity and causes fatigue. Lactic acid accumulation is also painful and is responsible for the 'burn' often referred to by trainers – to 'go for the burn' means to exercise to the limits of your anaerobic capacity. When exercise ceases, lactic acid must be cleared from the body in a process that consumes oxygen. The amount of oxygen required was termed the 'oxygen debt' by the British physiologist Archibald Hill. This is the reason that after a heavy game of squash you are still gasping for breath long after you have stopped dashing about. The more vigorously you exercise, the more lactic acid is produced, and the longer the period required for recovery. Anaerobic exercise therefore buys time, but only for a while, and at a price.

Although anaerobic metabolism releases energy rapidly, it produces relatively little ATP – only two molecules of ATP for every one of glucose. In contrast, aerobic metabolism is much more efficient and produces an additional 34 ATP molecules. Exercise that lasts for more than two or three minutes therefore relies increasingly on aerobic metabolism. About half of the energy needed to run the 1000 metres (2.5 minutes), 65 per cent of that required for running a four-minute mile, and almost all of the energy supply in a marathon comes from aerobic metabolism. Because aerobic metabolism is oxygen-dependent, the rate of ATP production is limited by the rate at which oxygen can be supplied to the tissues. In turn, this depends on the capabilities of the heart and lungs.

Oxygen Demands

At rest, an adult human consumes about a third of a litre of oxygen each minute. During strenuous exercise the oxygen requirement increases more than ten-fold in untrained people and as much as twenty-fold in top athletes. There must therefore be an enormous increase in the rate at which oxygen is taken up by the lungs and delivered to the tissues by the heart and circulatory system. Surprising as it may seem, the factor that limits oxygen uptake by the muscles is not the capacity of the lungs, or the ability of the muscle to extract oxygen from the blood; it is the rate at which the heart can pump blood around the body.

The normal output of the heart is 5.5 litres of blood a minute, which means that almost the entire volume of blood in the body (5 litres) is pumped through the heart each minute. In heavy exercise, the cardiac output can increase five-fold in normal people and in world-class endurance athletes it rises even more: they have maximum cardiac outputs of 35–40 litres per minute. In addition to ensuring that the skeletal muscles (which move the limbs) receive more blood, the increase in cardiac output is also important for extracting more oxygen from the air. Because blood flows through the lungs more quickly, it can pick up more oxygen each minute.

So how does the heart adjust its output to match the demands of working muscle? One way is by increasing the rate at which the heart beats, which is triggered by a rise in the circulating level of the hormone adrenaline. Another way is to increase the volume of blood pumped each beat. This is also stimulated by adrenaline, as well as by an additional mechanism discovered by the physiologists Otto Frank and Ernest Henry Starling and therefore known as the Frank–Starling effect. Their studies showed that if the heart muscle is stretched by returning blood it contracts more strongly, increasing the volume of blood ejected each beat. When the heart rate is increased, the blood circulates more rapidly, and consequently blood returning to the left chamber of the heart fills it more quickly and completely, which means that the force with which the heart contracts is increased. The amount of blood pumped each beat does not increase indefinitely, but reaches a maximum when exercise is only about a third of its peak capacity. Further increases in cardiac output are due entirely to an increase in heart rate.

In a closed system like the circulation, an increase in the force with which the heart pumps would lead to a rise in blood pressure unless there was also a fall in the resistance to the flow of blood. Pumping air into an empty bicycle tyre, for example, increases the tyre pressure if the inner tube is intact but not when it has a large slash in its side. Blood pressure does not rise during exercise because the resistance falls dramatically, due to a massive increase in blood flow to the muscles. In resting muscle, the finest blood vessels (the capillaries) are mostly closed. During exercise these dormant capillaries open up, so that the muscle is more thoroughly perfused and oxygen delivery is greatly increased. More oxygen is also removed from the blood: at rest, only around 25 per cent of the available oxygen is extracted by the muscles, but in heavy exercise this can rise to almost 85 per cent.

The increase in cardiac output may still not be sufficient to supply exercising muscle with the oxygen it needs, so during very heavy exercise blood is diverted away from less active organs to the muscles. The kidneys, for example, may get less than a quarter of their usual blood supply. In contrast, blood flow to the skin is usually maintained, or even increased, to help dissipate the extra heat produced by working muscles. More blood is also needed by the heart muscle, as patients with heart disease know only too well. They experience chest pain (angina) on exercise because their damaged coronary arteries cannot supply the heart muscle with the extra blood it needs. Only the blood supplied to the brain remains constant.

As everyone knows, you breathe faster and more deeply when you run, and the harder you exercise the greater the increase in breathing. Rapid changes in respiration occur within a few seconds of starting exercise, long before extra oxygen is needed by the muscles. It seems the body anticipates the coming oxygen demand and prepares itself in advance. If exercise is maintained, breathing increases further still. Physiologists are still trying to find out what triggers these changes in respiration. What is clear, however, is that breathing does not limit exercise – no one ever really 'runs out of breath'. In fact, most people tend to over-breathe during exercise. It may seem as if you are fighting for air, but the problem is not that your lungs cannot get enough oxygen but that your heart cannot deliver it fast enough to the tissues. Only at altitude does breathing limit performance.

Exercise can produce more benefits than physical fitness. It can also lift your mood. Chemicals known as endorphins flood the brain of the runner. Their name (*endo*genous m*orphine*) recognizes the fact that they interact with the same receptors as morphine. Like synthetic opiates they decrease pain, increase relaxation and make you feel good. Anyone feeling mildly fed up with life might be well advised to go out and do some physical exercise. Not only will the endorphins lift your mood, but your fitness will be increased and you will have the added benefit of feeling very virtuous about it all.

Although opiates such as morphine and opium are addictive, it seems unlikely that you can become physically addicted to endorphins, which are only present at low levels and have only mild effects. Nevertheless, many fitness fiends do develop some psychological dependence on the 'exercise high' and feel restless and irritable if they are prevented from exercising by injury or illness. Perhaps one should look on this as a blessing. After all, anything that induces us to take regular exercise is beneficial.

You Are What You Eat

Just as special high-octane fuels are used by Formula 1 racing cars, so diet is important for world records. Athletes burn calories with abandon. A top cyclist competing in the Tour de France uses almost 5900 kilocalories (kcal) a day, triathletes consume 4800 kcal, football players of league standard often expend 1500 kcal a day just in training, and running a marathon requires around 3400 kcal.[1] Other physically active workers, like lumberjacks, use similar amounts. In contrast, a couch potato only needs about 1500–2000 kcal daily (but often consumes many more).

Traditionally, athletes have been advised to eat a high protein diet. When I was an undergraduate, one of the attractions of being a member of the rowing crew was the special diet: steak for breakfast, lunch and dinner was thought to build muscle mass and enhance stamina. A recent survey of American college athletes revealed that 98 per cent of them also believed that a high protein diet improved performance. But this idea is misguided, for there is no scientific evidence that excessive protein consumption or expensive protein supplements have any beneficial effect on physical performance.

Carbohydrate intake is quite another matter. Many studies have shown that a high carbohydrate diet improves performance. For a physically active person, about 60 per cent of their calories should be supplied as carbohydrates and for those in hard training perhaps as much as 70 per cent. Glycogen, a carbohydrate stored in muscle and liver, is the main fuel supply for both anaerobic and aerobic metabolism, and the more intense the exercise, the greater the reliance on glycogen as a fuel. The glycogen stores in muscle are exhausted after about an hour of exercise, and if exercise is maintained for several hours those in liver also drop. This produces a slow decline in power because the athlete is forced to rely more heavily on fat, which cannot supply ATP at the same rate as carbohydrate.

During strenuous exercise, substantial depletion of the glycogen stores occurs and they must be refilled subsequently or the athlete will find they cannot exercise as much the following day. This means that after a hard day's training, bread and potatoes are better (unfortunately) than smoked salmon and cream cheese. Another problem is that even with an adequate intake of carbohydrate it takes a long time to rebuild the glycogen stores – at least twenty-four hours. Consequently, unless it is carefully managed, a hard training programme can lead to the gradual depletion of muscle glycogen stores over several days. This results in 'staleness', a condition in which the athlete suffers from increasing fatigue, because the energy available for exercise falls.

Endurance athletes sometimes use a technique known as carbohydrate loading to maximize the glycogen stored in their muscles before a competition. Trial and error has shown that the best way to do this is to first empty the glycogen stores in the relevant muscles by exhausting exercise. For a marathon runner, this might involve a 20-mile run. Further depletion is produced by eating a low carbohydrate diet for the next few days, while continuing training at moderate levels. Then, two or three days prior to competition, the diet is switched to one high in carbohydrates and exercise is reduced. This procedure of first emptying the glycogen stores and then eating a carbohydrate-rich diet results in an 'overfilling' of the glycogen stores in the muscles that have been worked. But it is really only useful for endurance athletes whose chosen sport involves high-intensity exercise for well over an hour. Anything less than that, and all that is needed is a normal, well-balanced diet.

Fat is an ideal storage fuel as it contains more energy, weight for weight, than carbohydrate. The potential energy stored in the fat of an average male student is an astonishing 95,000 kcals – more than enough to walk 9500 miles (or from Boston to San Francisco three times over). Women have relatively larger fat stores and can go even further. In contrast, the energy stored in carbohydrate reserves is only sufficient for a 20-mile walk. Obviously, marathon runners must rely on their fat stores to get to the finish. During moderately intense exercise, energy is derived in approximately equal amounts from fat and carbohydrates during the first hour, but after that the carbohydrate stores are gradually depleted and the body relies more and more on fats. Slimmers might do well to remember this fact.

Speed versus Stamina

It is difficult to excel in all sports. Sprinters and weight-lifters are not well adapted for endurance sports, while marathon runners can cover

How Muscle Contracts

The way in which muscle shortens has fascinated scientists for centuries. As recently as the 1950s it was suggested that muscle shortened because the contractile proteins themselves decreased in length. In other words, that the contractile proteins moved from an extended conformation to a shorter one, as rubber molecules do when elastic is stretched and then released, or like a slinky coil of wire contracts and extends as the coils are pushed together or pulled further apart.

It is now clear that these ideas were quite wrong. Muscle contraction results because two types of protein filaments slide past one another, so that the overall length of the muscle shortens without the proteins themselves doing so. A simple analogy is to touch the tips of your fingers together with your palms held at right angles to your fingers. If you now interlace your fingers, the distance between your palms shortens although your fingers (of course) remain the same length.

Contractile proteins come in two varieties: as thick and as thin filaments. The thick ones have many small hooks along their length that are able to attach to specific sites on the thin filaments. Breaking the connecting cross bridges, and then reattaching them to a new site a bit further along, results in the thin filaments being pulled in caterpillar fashion between the thick ones, so shortening the muscle. The more the filaments overlap, the more cross bridges can form, and the greater the force exerted by the muscle. Conversely, if the muscle is stretched so much that the filaments are pulled completely apart, no cross bridges can form and no force develops, so the muscle is completely relaxed.

Precisely how the cross bridges between the thick and thin filaments

26 miles at a rate of around one mile every five minutes yet cannot run a mile in under four. These differences result from both intrinsic genetic variation and the effects of training on the heart and skeletal muscles.

The muscles which we use to move our limbs are made up of many individual cells, known as muscle fibres. These fibres clump together to form long thin muscle bundles, which give meat its 'stringy' nature. In turn, the muscle bundles are assembled into muscles, that are attached to the skeleton by tendons. Muscle fibres come in two types, fast and slow. As their name implies, fast muscles contract rapidly. But they fatigue easily. They are used for short-term high-intensity exercise such as sprinting and weight-lifting, as well as in sports where short bursts of intense activity are encountered, like ice-hockey. Fast muscles run

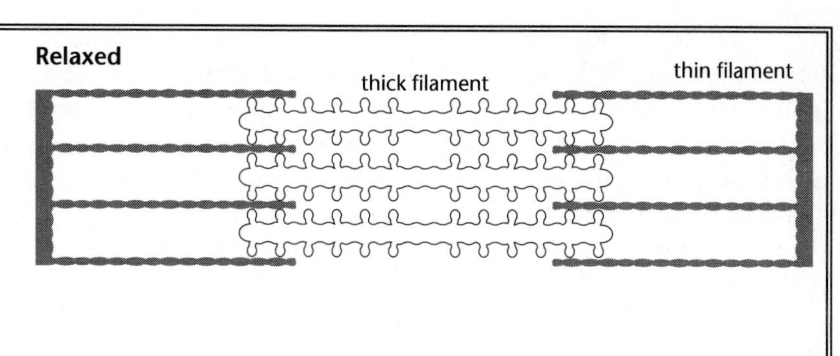

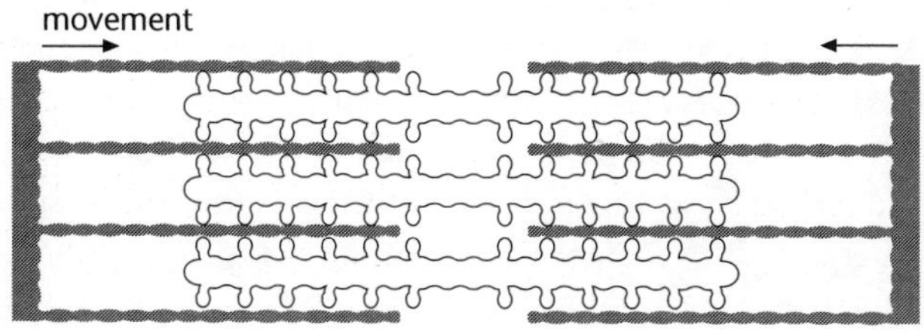

Muscle contraction

work is still not certain and it remains one of the great challenges for muscle physiologists. What is known, however, is that breaking and remaking the interfilament cross bridges is an energy-dependent process and consumes ATP. Rigor mortis occurs when ATP levels fall after death because ATP is actually required for the breaking of the cross bridges – in its absence the muscle is locked solid.

mainly on anaerobic metabolism, which does not need oxygen. Slow muscles contract at less than half the rate of fast ones but they are very resistant to fatigue. They are specialized for oxygen-requiring aerobic metabolism and are used in endurance sports like long-distance running and swimming.

About 50 per cent of muscle fibres in sedentary people are of the slow variety but in endurance athletes, such as cross-country skiers, it can be almost 90 per cent. Conversely, fast twitch fibres predominate in sprinters and weight-lifters. As might be expected, people who compete in middle-distance events or sports that require both speed and endurance (footballers, for example) have an equal percentage of fast and slow fibres. Thus, despite being found in non-athletes, an equal number of fast and slow fibres does not necessarily indicate slothfulness. Clearly, an individual who naturally has a predominance of fast twitch fibres is more suited to sprinting than to marathon running. A key question, therefore, is whether the relative number of each fibre type is determined solely genetically or if it can be modified by training. The current view is that in humans training has little effect on the distribution of fibre types – you are preprogrammed for speed or stamina by your genes.

Different muscle types are not confined to mammals. Pelagic fish like mackerel and tuna have slow muscles that they use for cruising around continuously at fairly low speeds, as well as fast muscle fibres that are used for short burst of speed – to escape a predator, for example. These two muscle types look very different, as you will observe if you examine a tuna fish in your local fish shop, or ask for *toro* and *maguro* in a Japanese sushi bar. Fast muscles are white in colour. Slow muscles are deep red because they contain large quantities of an oxygen-carrying molecule related to haemoglobin called myoglobin. This acts as a temporary oxygen store that is used during vigorous muscle contraction which squeezes the capillaries and reduces the flow of oxygenated blood. It is recharged during relaxation, when blood flow is restored.

Fast Forward

Even before the event, a sprinter's heart starts to race. As they crouch on the starting block, the suspense triggers a flood of adrenaline that kicks the heart into higher gear. Scientists have found that before a 60-yard sprint the pulse of a trained runner rose to 148 beats a minute, an extraordinary 75 per cent of the total increase in heart rate during the run. For a short dash, this anticipatory rise in heart rate is valuable, because it 'revs up' the body in preparation for the coming exercise. It is less useful for longer races, where a fast start is less important. Interestingly, it turns out that the longer the distance the athlete faces, the smaller the anticipatory rise in the heart rate. Does this mean that the tension (and thus the adrenaline level) is less before a longer race?

For the sprinter, a good start is essential. It provides those vital extra hundredths of a second that can mean the difference between success and failure. Move off too soon, however, and you may be disqualified for a false start. But how quick off the mark is too soon? Clearly,

Linford Christie on the starting block

it must be slower than the reaction time of the athlete – the time it takes for them to hear the crack of the starting pistol, for the nerve impulses to travel from the ear to the brain, be processed by the cerebral cortex and new signals sent to move the leg muscles. Human reaction times normally lie between 0.1 and 0.2 seconds, so the International Amateur Athletics Federation considers that any athlete who responds in less than 0.1 seconds has anticipated the starting pistol, and it is classed as a false start. At the 1996 Olympics in Atlanta, the British 100-metre sprinter Linford Christie set off 0.08 seconds after the gun, and was disqualified by the officials. But perhaps he should not have been. Recent studies suggest that in some circumstances human reaction times might be faster than 0.1 seconds. The physiologist Josep Valls-Solé and his colleagues found that the time that people took to move their wrist or foot in response to a flash of light could be almost halved if it was accompanied by a loud noise. They suggested that this 'startle response' bypassed the cerebral cortex and used shorter, and therefore faster, pathways in the brain. Interestingly, the subjects were aware that something different had happened – they felt that somehow they had moved without willing it to happen. It is possible that some top athletes can also access these pathways by 'psyching themselves up' on the starting block.

High-intensity exercise of short duration requires a very rapid supply of energy. At first, this energy comes almost entirely from existing stores of ATP and creatine phosphate, which are sufficient to sustain maximal exercise for about fifteen seconds. Subsequently, anaerobic metabolism is used to generate ATP from glycogen stores in the muscle. No oxygen is required for anaerobic metabolism and so far as their muscles are concerned, an athlete could run equally fast over 100 metres without breathing (and in fact some of them do so). But anaerobic metabolism generates lactic acid which accumulates in the muscle and contributes to fatigue. It is the gradual rise in lactic acid which accounts for the fact that a sprinter can run as fast over 200 metres as 100 metres, whereas they are significantly slower over 400 metres. Michael Johnson, who currently holds the world records for both events, clocked times of 19.23 and 43.18 seconds for the 200 and 400 metres, respectively – he would have had to have run the 400 metres in an impossible 38.46 seconds to maintain the same speed as in his 200-metre triumph.

The amount of the creatine phosphate stored in your muscles affects how long you can run at top speed, for only once it is exhausted does anaerobic metabolism kick in and lactic acid begin to build up. This can be critical for a top-level sprinter because even a few hundredths of a second can mean the difference between victory and defeat – between an Olympic gold and no medal at all. People who have naturally low levels of creatine are therefore at a competitive disadvantage. Creatine supplements can help even things out. The normal dietary intake is around one gram a day, but in vegetarians it is almost negligible because the main source of creatine in the diet is meat and fish. Consumption of 20 grams of pure creatine a day (far more acceptable than eating the fifteen steaks or so that would be needed otherwise) for a few days can significantly increase muscle creatine levels, improve performance in sprinters and enable more intensive training to be carried out. This practice does not contravene the International

Maurice Greene currently holds the world record for the 100-metre sprint (9.79 seconds). He also won gold medals in both the 100-and 200-metre sprints in the 1999 World Championship Games in Seville – the only man ever to have done so. Like all sprinters, he has powerful and well-developed muscles. His speed has not only won him titles; it also has more unusual benefits. While waiting at Seville airport, Greene observed a pickpocket steal a wallet from another athlete, and easily caught the culprit in the ensuing chase. It must have come as something of a shock to the thief to learn that his pursuer was the fastest man on Earth.

Olympic Committee doping regulations and there are no reported side-effects (as yet).

You have only to look at a world-class sprinter like Maurice Greene to see that their physique is very different from that of the distance runner. Speed is synonymous with strength, and sprinters have a well-developed musculature because large muscles are more powerful. It is obvious that the ability to explode out of the block and rapidly accelerate to top speed requires powerful leg muscles. A well-developed upper body is also essential, because when a sprinter runs, he pushes as hard as he can against the ground, first with one leg, and then the other. This tends to force his upper body to twist from side to side, which would impair his running. A strong upper body helps resist this force and keep the runner straight on the track.

Every runner must overcome air resistance. Running into a head wind is far more difficult than sprinting with the wind behind you. For this reason, a new world record requires that the speed of any following wind be less than 3 miles an hour. As much as 13 per cent of the energy costs of a sprinter goes in overcoming air resistance. In a middle-distance runner, it is around 8 per cent (because they go slower). Running behind another person virtually eliminates air resistance. Sprinters must keep in their lanes, but middle-distance runners are allowed to deviate. If you watch a bunch of them racing around the track you will observe that several of them lag behind the leader to take advantage of the slipstream, and then move out at the last minute to race ahead. Cyclists and racehorses use a similar strategy. This is particularly noticeable in team-cycling events, where a different member of the team will take the lead every few minutes. Gamesmanship, as well as physical ability, is the key to winning.

Staying the Distance

In the fifth century BC, the Persians invaded Greece. They landed at Marathon, a small town on the coast north of Athens. Their numbers were so great that when the Athenian army arrived it was heavily outnumbered, so messengers were sent to cities all over Greece asking for help. Herodotus relates that Phidippides, a trained long-distance runner, was sent to Sparta (150 miles away) where he arrived the day after leaving Athens. According to legend, he also ran the 25 miles[2] from Marathon to Athens a few days later to report the Greek victory over the Persians. But the traditional story is incorrect, for Phidippides was still in Sparta and it was actually another man, Eukles, who ran the first marathon. Perhaps he was not so experienced a runner as Phidippides, for having delivered his message he collapsed and died, thus ensuring that his feat would become immortal. Fortunately, few marathon runners die on the finishing line these days.

It is peculiarly apt that the winner of the marathon at the first modern Olympic Games, held in Athens in 1896, should also have been a Greek. This was a wonderfully amateur and good-humoured games in which the athletes were mostly self-selected. Thomas P. Curtis, an American who himself won the high hurdles, wrote that:

> 'On the last day of the games Greece came into her own. Loues, a Greek donkey boy, led all the other contestants home in the great marathon. As he came into the stretch, a hundred and twenty-five thousand people went into delirium. Thousands of white pigeons, which had been hidden in boxes under the seats, were released in all parts of the stadium. The handclapping was tremendous. Every reward which the ancient cities heaped on an Olympic victor, and a lot of new ones, were showered on the conqueror, and the games ended on this happy and thrilling note.'

From these auspicious beginnings, the marathon has evolved into a popular event in which both elite athletes and ordinary people test their stamina and courage. Each year, over 30,000 people run the London marathon and many more would do so if the numbers were not restricted. Many similar races take place throughout the world. But the marathon is not the ultimate challenge. There are longer races in more extreme environments, such as the Marathon des Sables, a gruelling 130 miles across the shifting sands of the Sahara in blistering heat, and a marathon that is run down the slopes of Everest with all the attendant problems of altitude. And then there is the Ironman triathlon, probably the most strenuous of all competitions, in which the athlete must first run a marathon, then cycle for 112 miles and finally finish with a 2-mile swim. The first triathlon was conducted in Hawaii in 1978 and there were only fourteen competitors. Like the marathon, however, the sport took off rapidly and today several million people compete in triathlon races of different distances

Haile Gebrselassie of Ethiopia winning the 1000 metres in August 1997. Like all long-distance runners, he is lean and wiry.

throughout the world. The top triathletes, like their decathlon colleagues, are an elite fraternity, for they achieve world-class performances in more than one sport.

A marathon is a test of endurance. The current world record, held by Ronalda da Costa of Brazil, is two hours, six minutes and five seconds. This is the equivalent of a 4.8-minute mile, a pace far faster than most untrained people can run over a single mile. Most people (even those who are trained) take far longer. The average time in the London marathon, for example, is between three and four hours.

A fast starting pace is not so critical in a marathon. What matters most is whether you can keep up a steady pace for the whole course. During a long-distance race, almost all energy is derived from aerobic metabolism so the runner must maintain a speed that allows oxygen to be supplied to the muscles at the same rate as it is consumed. Consequently, the pace is slower than that of a sprint. However, the very low level of anaerobic metabolism minimizes lactic acid accumulation and enables the distance runner to travel further. Slow muscle fibres, which are specialized for aerobic metabolism, are mainly used for endurance running.

Long-distance runners are lean and light: a 3:1 ratio of height (in centimetres) to weight (in kilograms) is supposed to be ideal. They have a mere 3 per cent body fat, less even than gymnasts and professional footballers and substantially less than sedentary people (who average around 15 per cent). This reduces the amount of 'dead weight' they have to carry and helps them keep cool during a long run. Overheating is a significant problem for a long-distance runner which is why athletes douse themselves with water and take drinks constantly during a race, and why in hot climates marathons are scheduled during the cool of the early morning.

For the first hour and a half of the race, energy is derived from glycogen stored in the muscles. Once these stores have been depleted the runner relies more and more on fat as a fuel. More oxygen is required to metabolize fat than carbohydrate, so the oxygen demand increases after the glycogen stores are used up. Around the 15–20-mile mark, most people suddenly feel tired and breathless, the low blood sugar level makes them dizzy and nauseous and they have to slow down. They have 'hit the wall'. Mike Stroud describes it thus:

> 'All pleasure had departed. My mind and body were hurting and my legs had become a strange mixture of rigidity and floppiness . . . they had become quite wayward and beyond my control . . . I was barely able to go on running and kept tripping over my own feet.'

A friend of mine who 'hit the wall' while cycling thought that his brakes had suddenly jammed on. He got off to see what was the problem, only to discover it was not his bike that was at fault, but his body! The change-over from running on carbohydrate to running on fat is highly unpleasant and things do not get much better even if you persevere at a slower pace. The next few miles are fearfully hard work.

Not so the last one. The excitement of finding yourself close to the end floods the body with adrenaline, providing a final boost that helps carry the novice – and even the experienced athlete – right to the finish.

Running Down

Some people claim to feel tired at the very thought of exercise, but fatigue is a real physiological phenomenon. It is the failure of muscle to maintain its power output during a prolonged contraction or a series of repeated contractions. This is what causes your arm to flag in an arm wrestle, accounts for your inability to perform repeated press-ups (or none at all in my case) and limits your capacity to run fast for long distances.

Fatigue can result from changes in the muscle cells themselves. One obvious mechanism that would produce a loss of power would be a failure to balance the energy (i.e. ATP) consumed by contracting muscle with the rate of energy production. But although ATP levels do fall in very intense exercise, they are never totally obliterated. Muscle cells in which ATP levels drop to zero develop rigor, the muscle contracture that causes stiffening of the body after death. Even during the most intense exercise, rigor is never observed in life. Perhaps fatigue should therefore be considered a protective mechanism, that forces muscles to stop long before ATP drops to a level that threatens their survival.

So what does cause muscle fatigue? There seem to be two main mechanisms, both of which involve the calcium ions that trigger muscle contraction. In response to a prolonged contraction, the amount of calcium that is released from intracellular stores in the muscle gradually falls, so that contraction is stimulated less effectively. A different mechanism seems to be responsible for the fatigue produced by repeated short contractions. In this case, the muscle stores seem to grow tired of releasing calcium. Why this should be the case is not entirely clear, but it is thought to be related to the accumulation of metabolic breakdown products that occurs during intense activity. These also inhibit the strength with which the contractile proteins can produce force.

Depletion of muscle glycogen is the main cause of exhaustion in endurance events; this is what saps your strength and makes your legs feel like lead. Fat metabolism is unable to provide ATP at the same rate as oxidation of muscle glycogen.

A rise in body temperature can also cause fatigue. In a short sprint, the amount of heat generated by working muscles can easily be disposed of, but in sustained exercise it can be more of a difficulty, especially in a hot climate. Every year, several runners in the London marathon collapse with heat exhaustion. The problem arises because there is a conflict between the demands of muscle and those of heat loss – blood that is directed to the skin for cooling purposes cannot be used to supply the muscles with oxygen. A failure of thermoregulation may explain why fatigue occurs more rapidly when exercising in a hot environment than under cold conditions. It is not so much a lack of fuel, but a signal originating in the brain that tells you to slow down or stop to prevent overheating. This mechanism appears to switch on when the body temperature rises above about 40 °C.

Finally, fatigue and muscle weakness also result from tissue damage. Overstretched muscle becomes inflamed and swollen, which limits its capacity to generate force. It can also be very painful. This type of muscle damage accounts for the stiffness that follows an unaccustomed bout of exercise, and it takes several days to recover from. Even fit individuals who take unaccustomed types of exercise can develop soreness, as happens with many people who ride a horse for the first time.

Working Out

One warm summer morning, I had to take the coach to London. As usual, I left things to the last minute and when I rounded the corner I saw the bus was already standing at the stop, about a hundred metres away. Because there was a queue of people getting on board I decided to sprint for it. I raced along the pavement, my chest heaving as I fought for oxygen, my heart thumping, my temperature climbing so rapidly I seemed to be steaming. Muscles unaccustomed to the exercise started to protest and shooting pains knifed into my side as lactic acid seared my diaphragm. By the time I reached the bus I was in a state of near collapse, gasping for breath, my muscles quivering like jelly, drenched in sweat and feeling sick. I was a good way to London before my heart stopped racing, my breathing returned to normal, my calf muscles unknotted and I finally cooled down. Two years before, when I went to the gym three times a week, I could have sprinted the same distance with comparative ease. Sitting on the bus, I felt as if I had run a marathon. So what makes the difference between being fit and unfit? And how does training prime the body for speed and increase stamina?

One of the more immediate benefits of training is the improvement of muscular coordination. When we walk, only some of the individual muscle fibre bundles in our muscles actually contract. When we run, more and more of them are recruited into action. For maximum efficiency, muscle fibre bundles must contract simultaneously. Synchronization of the muscle bundles occurs quite quickly with training, producing a rapid improvement in speed and strength. This is the main reason why it seems much easier to cycle uphill after only a week or two of daily practice. Even with training, the muscle bundles never all contract simultaneously. Were they to do so, the force generated might be close to the limit at which bone breaks. Total synchronization of muscle fibre contraction may perhaps explain the extraordinary power that athletes – and even ordinary mortals – can sometimes generate under extreme stress. Stories of people lifting a car off an accident victim, or of athletes suddenly producing a never-to-be-repeated performance that far outstrips their personal best, are not uncommon. Such synchronization may also have devastating consequences. In 1995, one of the contestants for the World's Strongest Man title generated such force in his arm muscles during the arm-wrestling competition that he snapped his own arm bone.

Practice also perfects skilled movements and improves judgement. A javelin thrower must judge when to let go of the javelin, a long-jumper when to take off, and a tennis player must learn how to place the ball out of his opponent's reach.

Training delays the onset of fatigue and improves the strength and power of muscle. This is principally a result of changes in the heart and skeletal muscles that improve the delivery of oxygen to the muscles and increase the efficiency of energy production. Improvements in these processes can be attained with even a relatively modest training programme. The length of time you can run before stopping from exhaustion, for example, more than doubles after only three to four weeks of regular exercise, and endurance improves even more markedly with intensive training. Sprint performance is also enhanced by training but this is mainly due to the ability to run fast for longer, rather than an improvement in absolute speed.

The effect of training on the heart can be dramatic. A trained Olympic cross-country skier has a maximum cardiac output more than double that of a healthy but sedentary person of the same age. The peak heart rate does not change with training. Rather, it is an increase in the volume of blood that the heart can pump each beat (the stroke volume) that enables trained athletes to pump more blood each minute than their untrained counterparts. Echocardiography, a technique in which sound waves are used to measure the size of the heart, reveals that this is because marathon runners have larger hearts. Regular aerobic exercise also increases the size of the heart in ordinary people.

Although training has no effect on the maximum heart rate, it slows the resting pulse. This is because the increased stroke volume means that the heart has to beat less often to supply the same amount of blood. The heart rate of an untrained person is 70 beats a minute, whereas that of a top athlete may be as low as 40 or 50 beats a minute. Even minimal training can lower the resting pulse – five minutes of rope skipping every day for a month is enough. The great advantage of a low resting heart rate is that there is a greater margin before the maximal heart rate – about 200 beats a minute in both trained and untrained people – is reached. This gives trained athletes a much greater peak cardiac output so they can deliver considerably more oxygen to their muscles.

The skeletal muscles are also affected by training. In particular, their ability to produce the high-energy molecule ATP is increased. Glycogen stores enlarge and the efficiency of metabolism improves. Slow muscle fibres – used in endurance sports – develop increased numbers of mitochondria, the organelles that make ATP, and their capacity to use fat as a fuel improves. In the fast muscle fibres that are used for sprinting the amount of lactic acid produced by a given work rate falls and a higher lactate concentration can be tolerated without discomfort. Blood flow to both types of muscle is augmented and capillary density rises, which improves the muscle oxygen supply. Muscle mass increases because the individual fibres become larger, augmenting muscle strength. These changes are strictly local, being confined to the muscles used during training. As a student, my Cambridge college was a 3-mile bicycle ride from the centre of town and we endured constant teasing that the enforced daily exercise swelled our calf muscles to giant proportions. There is only limited truth in this idea, however (observation was not the teasers' metier), for endurance training produces only modest increases in muscle mass. Specific exercises are required to achieve the Herculean proportions of a Charles Atlas.

Unfortunately, the effects of training are not permanent. The heart rate returns to its previous level within a few weeks of stopping regular exercise. It is far harder to get fit than it is to lose it, for what took a month to gain can be lost in a mere week. This is not an excuse for not getting fit, but rather a spur to not slacking (or so I tell myself).

The Ultimate Limits

While training can impove individual performance, ultimately physical capacity is determined by one's genes. The genes that influence physical performance are just beginning to be discovered. The first report of such a gene was published in the journal *Nature* in 1998. This gene encodes a protein called angiotensin converting enzyme (ACE) which is important in the regulation of the circulatory system. Everyone has two copies of a given gene, one from each parent. What the scientists found was that army recruits who had two copies of a specific variety (*I*) of the ACE gene were able to lift weights for eleven times longer than those who had two copies of the *D* variety of the gene. Men who had one copy of each variety lasted for an intermediate time. Interestingly, this difference only showed up after ten weeks of physical training – no difference in the ability of the recruits was observed before training. High-altitude mountaineers who had routinely climbed over 7000 metres without the assistance of oxygen also had at least one copy of the *I* variety of the ACE gene. The *I* variety of the gene is associated with much higher activity of the angiotensin converting enzyme, but why this should improve performance after training is still not clear.

Ultimately, speed and stamina must be limited by the physical properties of the muscles and cardiovascular system. The rate and force at which the heart and skeletal muscles can contract have very definite physiological limits. The maximum heart rate in a fit young person is around 200 beats per minute, regardless of training.[3] This limit is set by the fact it takes a finite time for the heart to refill. It is obviously highly inefficient for the heart to contract before it is full. Indeed, it can be fatal. Ventricular fibrillation is a condition in which the heart beats uncontrollably fast in an asynchronous fashion. This prevents refilling of the great chambers of the heart and, unless the heart can be shocked back into its normal rhythm, inevitably leads to death. The maximum amount of blood that the heart can pump each beat is also limited – by the size of the heart. Larger hearts make better athletes and one of the main benefits of regular exercise is that it increases the size of the heart.

The maximal force that a skeletal muscle can exert appears to be around 4 to 5 kg force per square centimetre of cross-sectional area (kgf/cm^2). In general, therefore, strength is achieved by increasing muscle mass – the fatter the muscle, the greater the power output. Some invertebrate muscles, however, can do better than human ones. Bivalve molluscs like cockles and mussels protect themselves from predators, or the receding tide, by closing their shells. The adductor muscle that closes the shell can exert a maximum force of 10–14 kgf/cm^2, two to three times that of a mammalian muscle. Moreover, bivalves can stay closed for many hours, because the muscle possesses a unique 'catch' mechanism that allows it to remain contracted without consuming ATP. To try and pull the shell apart, as starfish do, is very difficult. In the tug of war between starfish and shellfish, the bivalve is almost always the winner. Its muscle has the greatest endurance.

Finally, as in all walks of life, a key difference between those who succeed and those who do not is motivation. The ability to push yourself to the limit, and to keep a cool head while doing so, is the hallmark of the champion.

Gender Issues

With the exception of long-distance swimming, women lag behind men in strength, speed and stamina in almost all sports. The reason for this is far from clear. Part of it must be simply a matter of training and opportunity.[4] It is very noticeable from viewing old film footage that even world-class women tennis players of twenty years ago were not as fast nor as hard-hitting as those of today. Female athletes have also been closing the gap steadily in track and field events and their world records more nearly match those of men. But women are still not as fast, nor do they have as much stamina. The fastest women's time over 100 metres is

Making a Splash

Swimming uses about four times the amount of energy that running the same distance does. This is partly because drag produced by friction with the water is a significant factor, whereas air resistance is rarely a problem for runners. Competition swimmers shave their body hair to reduce their drag. Wearing a wetsuit allows you to go even faster, because drag on the body is reduced further still.

In swimming, power is provided by the arms and the legs are largely irrelevant. This is mirrored in the types of fibre found in the muscles – swimmers have a greater proportion of slow muscles in their arms than in their thighs. The leg kick in front crawl is to produce a streamlined shape in the water, rather than to supply motive force, as you will find if you simply kick your legs without moving your arms. Swimming with only your arms is very tiring, however, because your legs sink and impede your movement through the water.

Keiron Perkins, Olympic gold medallist in Atlanta, 1996

10.49 seconds, significantly slower than the men's record of 9.79 seconds. Over a marathon, the gap widens further, for the women's record is over fourteen minutes longer than the men's. So the question remains: why are women slower and will they ever catch up with men?

In sports where physical attributes like strength and speed are less important, women compete on equal terms with their male counterparts; in showjumping, for example. This suggests that it is the physical abilities of women that account for the differences in track and field events, rather than the fact that they are less competitive, less aggressive or less determined. And there are well-documented physical differences between men and women (in addition to the obvious ones). Among world-class cross-country skiers, the maximum oxygen uptake for women is only 43 per cent of that of their male colleagues. Even when differences in body weight are taken into account, it is still 15–20 per cent lower. In part, this is because women have a higher percentage of body fat than men and relatively less muscle. Indeed, some studies suggest that if differences in muscle mass are taken into account, women would have as high a rate of oxygen uptake as men. But men have a further advantage in that they have more haemoglobin than women (10–14 per cent), which improves the oxygen-carrying ability of their blood. Women are also smaller than men and they have correspondingly smaller hearts, so that the volume of blood pumped each beat is usually 25 per cent less than that of men. Because endurance is limited by cardiac output, this suggests that women will have less staying power in long races.

The fact that men have greater amounts of the male sex hormone testosterone may account for some of these physical differences, and help explain why women still cannot match the men's world records. It is noteworthy that several of the women's world records were set by athletes who either later confessed to taking anabolic steroids (which mimic the action of testosterone on muscle mass) or are widely suspected of having done so.

There is one event, however, in which women surpass men and that is long-distance swimming. Again, this can be explained by their different physiologies. Fat is less dense than water and tends to float, while muscle is heavier and sinks. Thus, because women have more subcutaneous fat, they float more easily than men. This means that women swimmers expend less energy in moving through the water than men and their legs are also closer to the surface, giving them a more streamlined shape. This helps explain why the world records for women more closely approach those of men in speed swimming than they do in athletics. In long-distance swimming the woman has a further advantage, because her fat affords greater insulation. Indeed, the current record for swimming the English Channel (21.5 miles) of seven hours and forty minutes is held by a woman. The men's record does not approach it, being eight hours and twelve minutes.

Enhancing Performance

The use of performance-enhancing drugs has its origins in antiquity. At the time of the Crusades, the Ismaili Muslims sent their warriors into battle, or on murder errands, high on hashish. Their ferocity and fearlessness are commemorated in the word 'assassin' which derives from the Arabic *hasisi*, meaning hashish-eater. In the nineteenth century, the British navy issued its sailors with a daily tot of rum to 'stiffen up the sinews' for battle. And the conditions in the Vietnam War were so horrendous that many US soldiers turned to drugs such as marijuana, cocaine and heroin. All these drugs were in one sense performance-enhancing, because they helped allay fear in what was a highly dangerous situation. Some, like cocaine, were also stimulants that helped overcome fatigue and injury (South American Indians have chewed coca leaves for centuries to subdue hunger and improve endurance). But none of them enhanced muscle mass or strength.

During the nineteenth century, drug use by athletes became commonplace. Caffeine, alcohol, cocaine, opium, ether, heroin, digitalis and even strychnine (a poison) were consumed in the hope that they would improve performance. Not unsurprisingly, fatalities followed. An English cyclist, who overdosed on tri-methyl during a race from Bordeaux to Paris in 1886, has the questionable distinction of being the first athlete to die from using a performance-enhancing drug.

As our understanding of human physiology developed and winning at sport – rather than 'playing the game' – became increasingly important, so athletes experimented with an ever-increasing range of drugs. Testosterone and synthetic anabolic steroids were introduced in the early 1950s when it was discovered that they build muscle mass. By the mid-1960s their use was widespread among weight-lifters and shot-putters, and by the late-1960s they were also being taken by runners. In 1967, the International Olympic Committee (IOC) decided to call a halt. They passed rules prohibiting the use of performance-enhancing drugs and instituted random drug tests. Currently, over 100 substances are banned by the IOC.

The increasing commercialization of sport, with sponsorship and big prize money accruing only to the most successful, places a particular premium on winning. Add to that the relatively short professional

The Olympic Ideal

We may sing of no contest greater than Olympia
just as water is the most precious of all elements,
just as gold is the most valuable of all goods,
and just as the sun shines brighter than any other star
so shines Olympia, putting all other games in the shade.

PINDAR, First Olympic Ode

The first recorded Olympic games were held in 776 BC and were a purely local affair that lasted only one day. They began with sacrifices to Zeus in the morning and concluded with a single foot-race in the afternoon, which was won by Coroebus of Elis – the first Olympic victor. By 650 BC the games had grown much larger. Citizens of many cities came to compete, including some from Italy and Asia Minor. The number of events had expanded to include several foot-races of different lengths (including one of around 5000 metres), boxing, chariot racing, horse-riding, pankration and the pentathlon (running, jumping, discus, javelin and wrestling). There was also one particularly gruelling event in which contestants raced in full armour (which weighed around 110–130 kilos) for 768 metres – a reminder of the importance of athletics in ancient Greece as a training for war. The winners received a wreath of olive leaves and very little else. But they brought fame and glory to their homeland, much as modern Olympic athletes do.

Although the first Olympics are often held up as an ideal of excellence and fair competition, this was not really the case. Just like today, they were plagued by politics and commercialism. Nor did their athletes refrain from cheating, although it took the form of bribery rather than drug-taking.

Greek black-figure amphora depicting foot-racing, dating from the 6th to 5th century BC. Panathenaic amphoras like this one were used to hold the oil given as a prize at the quadrennial games in Athens.

life of the athlete and it is easy to see why a number of athletes break the rules and experiment with performance-enhancing drugs. The greater the number who do so, the harder it is for the others to resist. As one athlete commented, 'If you're not on anything, it's like lining up on the blocks with trainers when everyone else is wearing spikes.' But drugs are not illegal simply because they are considered 'unfair'. They are banned because most of them have serious side-effects. It is somewhat ironic that athletes go to great extremes to improve their fitness and then abuse their body with drugs that can produce sterility, liver cancer and sudden death from heart failure.

The most notorious of the performance-enhancing drugs are the anabolic steroids, which are synthetic analogues of the male sex hormone testosterone. These drugs increase muscle mass and strength and are taken to improve performance in sports that require strength, speed or power, such as weight-lifting, running and swimming. They are also used by body-builders. Because anabolic steroids are most effective during training, their use can be stopped three to four weeks before competition, allowing the drugs to wash out of the system and the athlete to be clean if tested after the event.

It is now incontrovertible that anabolic steroids enhance speed and stamina. The best evidence comes from records kept by doctors and coaches of the former East German state, who ran a well-orchestrated state doping campaign for their star athletes for many years. As a consequence, East Germany dominated women's swimming between 1973 and 1989, winning eleven out of thirteen medals at both the 1976 and 1980 Olympics, and ten out of fifteen titles at the 1988 Olympics. Petra Schneider, for example, set a new world record in the 400 metres individual medley at the 1980 Moscow Olympics that held for an astonishing fifteen years. She later revealed she had unwittingly taken anabolic steroids, which presumably contributed to her extraordinarily fast record.[5]

Unfortunately, anabolic steroids have many side-effects, which include an increased risk of heart disease, liver cancer, kidney damage and personality disorders. In male athletes, they may lead to a down-regulation of endogenous testosterone levels that persists even after consumption of anabolic steroids has ceased. Shrivelled testes and infertility are often the result. Female athletes suffer masculinizing effects, including altered menstruation, increases in body hair, and disruption of normal growth. Christiane Knacke-Sommer, the first woman to swim the 100 metres butterfly in under a minute, and several of her colleagues, were given steroid hormone pills by their East German trainers. They had little choice in the matter. As Knacke-Sommer told a Berlin court, if they did not take the 'supplements' or 'vitamin pills' they were out of the team. A number of them are now paying a heavy price for doing so, for their health has been permanently damaged – so much so, in fact, that several former East German swimming coaches and doctors have been convicted of causing them harm.

The East Germans are not alone. In 1988, Ben Johnson was stripped of his gold medal and and banned for life from professional athletics when he tested positive for anabolic steroids after winning the 100-metres race in the Seoul Olympics in a record time of 9.79 seconds. This event was somewhat of a watershed in the public perception of drug-taking by athletes. Prior to Seoul, the media often ignored the issue, even when it was pointed out to them. When Johnson tested positive, however, drug-taking became headline news overnight and since then it has rarely been out of the news.

Perhaps Johnson was lucky to have been caught, for high doses of anabolic steroids can damage the heart. The sprinter Florence Griffith-Joyner, affectionately known as Flo-Jo, died of a heart seizure at the tragically young age of thirty-eight. She won three Olympic golds in 1988 and set world records for both the 100 metres (10.49 seconds) and 200 metres (21.34 seconds) that have not been surpassed. She was graceful, beautiful, had outrageously long fingernails and wore flamboyant outfits on the track. She also had a muscular physique, a deep contralto voice and, although it was never proven, was widely suspected of taking anabolic steroids.

Anabolic steroids are not the only drugs used by athletes to improve performance. Growth hormone, amphetamines, adrenaline, erythropoietin, and a host of other less well-known drugs are also used. Growth hormone is given to undersized children when they are young to help them achieve a normal height. It stimulates bone and muscle growth, and decreases body fat. It is particularly attractive to the athlete because there is no foolproof way to distinguish synthetic human growth hormone from that produced by the body. And now that human growth hormone can be made by bacteria in large quantities it is far less expensive and more easily obtainable. But it is not without its risks. An excess of growth hormone in adults causes a condition known as acromegaly in which the hands, feet and facial bones grow very large.

Amphetamines are also sometimes taken by athletes. Their common name of 'pep pills' explains why they are popular. They produce arousal, reduce fatigue and pain and generally put the body in a 'hyper' condition by stimulating cardiac output, increasing the pulse and breathing rates and raising the level of blood sugar. Amphetamines mimic the action of the natural hormone adrenaline that prepares the body for fight or flight. Indeed, adrenaline is itself sometimes taken by athletes. But amphetamines also have negative effects such as dizziness, agitation and confusion, and are ill advised for sports where judgement, concentration and a 'cool head' are required.

In the summer of 1998, the Tour de France was racked by scandal. It began when a masseur for the Festina team was stopped at the Franco-Belgian border and a range of drugs was discovered in his car. Subsequently, five of the team admitted taking the drugs. Despite considerable protest from the cyclists, tests were carried out on members of several other teams and their luggage was also searched for drugs, with positive results. Eventually, more than eighty of the 189 competitors were either disqualified because of drug-taking or dropped out of the race. The drug that most of them were taking was human erythropoietin, a hormone that stimulates the production of red blood cells (see Chapter One). Injecting erythropoietin is simply a more sophisticated form of blood doping, a much older practice in which a blood transfusion is given before competition to increase the athlete's number of red blood cells and thus their oxygen-carrying capacity. It is still not proven that this enhances performance, although many athletes believe that it does. The worry is that the greater viscosity (thickness) of the blood will precipitate blood clotting and increase the risk of stroke and heart attacks.

What about the more common stimulants, like coffee and alcohol, that most of us consume in great quantities throughout our lives? Perhaps surprisingly, caffeine actually seems to increase performance. One study found that taking the equivalent of two and a half cups of strong coffee an hour before exercising significantly increased endurance. Those people who drank coffee were able to exercise for more than ninety minutes while those who drank decaffeinated coffee managed only seventy-five minutes. The coffee drinkers also felt less exhausted. Quite how caffeine produces this effect is not certain but it seems that the drug may facilitate the use of fat as fuel (and spare the body's limited carbohydrate reserves) and that it also has a direct action on the muscle itself. The IOC sets a limit of 12 micrograms of caffeine per millilitre of urine. To achieve this level, an athlete would need to drink between six and eight cups of coffee at one go and be tested within two hours. Unlimited caffeine use, of course, has adverse effects. It can cause headaches, fine tremor and extra heartbeats. And it acts as a potent diuretic which could be a problem in a long run – not just because of the need to relieve yourself but also because enhanced fluid loss can cause dehydration.

The benefits of alcohol are thought to be largely psychological – to allay nerves and increase self-confidence. It also reduces fine tremor, which may be of value to athletes who need a steady hand. However, its use is illegal and in the 1968 Olympics two pistol shooters were disqualified for taking alcohol before competing. Too much alcohol, as many people know from experience, is of course detrimental to performance.

Animal Magic

Training helps improve performance but there must be some limit to how fast or how far a human can run, or how high they can jump. What are these physical limits? And how do they compare with those of animals? It is not easy to answer these questions, because records are constantly being broken. Elite athletes, improved training, better shoes, better equipment, the right track, a following wind, all undoubtedly contribute. Nevertheless, world records are rarely exceeded by large amounts and it is most unlikely that one day a man will appear who can match the speed of the cheetah. We are therefore probably safe to assume that the current world records are not far off the limits for humans.

A top sprinter can run 200 metres at a rate of 22 miles an hour and her long-distance counterpart can cover a mile at 15 mph. Although this is far faster than most people can run, it pales into insignificance beside what other animals can achieve. A whippet races at 35 mph, jack rabbits can sprint at 40 mph, a red fox lopes along at 45 mph, antelopes have been clocked at 60 mph, and the cheetah can reach an astonishing top speed of 70 mph. Even the ostrich – which like humans has only two legs – can run at an impressive 35 mph. Animals also win in the endurance stakes. A horse, for example, can gallop at 15 mph for 35 miles, camels can cover 115 miles in twelve hours and a red fox being chased by hounds was recorded as running 150 miles in a day and a half. Speed and endurance are important for both predator and prey, but predators tend to sprint faster while their prey often have superior endurance and agility.

Both the length and the frequency of the stride are important for speed. The beautiful hypnotic 'slow-motion' gait of the giraffe results because it has a long stride but a slow pace. Smaller animals can achieve similar speeds with a shorter stride if they move their legs faster, like the warthog does. A similar comparison can be made when sitting in a street café and watching the passers-by. People with short strides often have to jog to keep up with their longer-striding companions. The fastest runners combine a long stride with a fast pace.

Animals that run fast tend to have long legs in relation to their size, which gives them a long stride. Many have evolved longer legs by modifying the bones of the feet. Carnivores and birds tend to run on what corresponds to the ball of the foot. This adaptation is taken even further in hoofed animals where the bones of the feet are fused together for strength to create a hoof. The horse is left with a single digit and, in effect, runs on the tips of its toes. Fast animals also lighten their limbs, by reducing the size of the bones in their feet and moving their muscles, and as much of the other tissues as possible, closer to the body. Long lean legs are the hallmark of the runner. The flexible spine of cats and dogs adds further length to their stride. When its back is extended, a cheetah is actually several inches longer. It must time the flexing of its spine so that its back is only extended when the back legs are pushing against the ground.

Fast runners must also move their legs rapidly. At full gallop, a horse manages 2.5 strides a second, and a cheetah at least 3.5. But the faster the rate of stride, the faster the leg muscles must contract. Ultimately then, the limit to speed must be determined by the rate of muscle contraction. This is roughly the same for all mammalian muscle fibres. However, longer muscles contract more slowly, which means that in large animals the advantages of long legs are offset by the slower rate of stride. This is one reason why the giraffe, despite its much longer legs, cannot compete with the cheetah. Some animals, like horses, get around this problem by having relatively short muscles and long tendons.

The cheetah is a sprinter *par excellence.* The fastest animal on Earth, it has a top speed of around 110 kph (70 mph). Even more remarkably, it takes only three seconds to reach this speed. But it cannot keep up this fast pace for long. Most chases are limited to less than half a minute because the intense anaerobic exercise builds up a large oxygen debt and causes a steep rise in body temperature (to almost 41°C, close to the lethal limit). A long recovery period must follow. This high energy expenditure means that the cheetah must chose its prey carefully, for it cannot afford too many unsuccessful chases.

The site at which the muscle tendon attaches to the leg bones also affects the rate at which an animal can run. In fast runners, the muscle is attached close to the shoulder joint, which means that less energy is needed to move the limb. In effect, these animals operate in high gear all their lives. Walking animals (like humans) and digging animals (like badgers) work in low gear. Their muscles are attached further away from the shoulder joint, which gives them more power but less speed. Another trick used by fast-running animals is to use several muscles to move different joints of the leg forward simultaneously. This speeds up the foot in a similar fashion to the way in which a man's speed is enhanced if he walks up a moving escalator. The more joints that can be moved together, the greater the velocity of the leg. By running on their toes, horses acquire an extra joint and hence more speed.

Some animals use elastic recoil to help propel themselves forwards. A ligament in the foot of the horse stores energy when the foot touches the ground and releases it again when the foot takes off once more.

When the foot impacts the ground, the fetlock joint bends and in doing so stretches an elastic ligament that wraps around the bent joint. When the foot leaves the ground, the joint straightens and the ligament snaps back to its original length, releasing the stored energy and giving the leg an upward boost. The elastic ligament reduces the need for a heavier muscle and the lighter limb facilitates speed. As a consequence, the horse is an extremely efficient runner.

The long Achilles tendons of the kangaroo serve a similar function to those of the horse. They save as much as 40 per cent of the energy cost of jumping on its hind legs and enable the kangaroo to increase its speeding of hopping from 7 to 22 kilometres per hour without using any more oxygen. In other words, it takes no more effort for the kangaroo to go faster! This is because the animal uses its tendons like the springs of a pogo stick to help it bounce along. Like a bouncing ball, the kangaroo uses most energy in its first hop and subsequent hops are aided by elastic rebound. More energy is saved by elastic storage at high speeds, so relatively less work is needed.

A simple experiment illustrates the importance of elastic recoil in saving energy. Put this book to one side and stand up and rapidly do a series of ten deep knee-bends. Then repeat the knee-bends, but this time count to sixty before you straighten your legs. You will find that the exercise now becomes far more strenuous. The reason is that the extensor muscles are tensed during squatting to control the rate of downward movement. If they shorten again immediately, the tension in the muscle provides an elastic rebound but if the tension is allowed to decay, there is no elasticity to help. Elastic recoil in your muscles helps you to bounce up and down with relative ease. It also puts the spring in your step and helps you save energy when running. Energy is stored in the calf muscle of the leg and the Achilles tendon when the foot contacts the ground, and is released again almost immediately when the foot pushes off the ground and the muscles shorten. Running shoes are designed to help amplify this elastic rebound.

'Four legs good, two legs bad' was the famous maxim of the animals in George Orwell's satire *Animal Farm*. It is certainly true that both speed and endurance records are held by quadrupeds, but are four legs really better than two? Unfortunately, the answer to this question is not straightforward because it is not simply the number of legs that determines speed; the size of the animal, the length of its leg, the flexibility of its back, and its gait all make important contributions.

Size Matters

As always, size matters. It becomes increasingly difficult for animals to run as they get larger. This is because the force that a muscle can exert increases as the square of its cross-section. However, the mass of an animal increases as the cube of its length. Double the length of an animal and its weight increases eight-fold but the ability of its muscles to generate force only increases four-fold. As animals get larger, therefore, it becomes increasingly difficult for them to move their limbs. If they grow really large, they may find it hard to support their body

The American photographer Eadweard Muybridge was one of the first to explore how humans and other animals run. In the 1870s, he set up a line of twenty-four still cameras at Leland Stanford's private racetrack in Palo Alto, California, and took successive snapshots as a horse galloped past. His photographs resolved the controversy about whether or not a horse has all four feet off the ground during the gallop. It turns out the answer is yes – for one-quarter of a stride, the horse is suspended in the air. But this occurs when the horse's legs are tucked up underneath its stomach, not when they are extended, as had been thought previously and as many artists had portrayed them.

even when they are not moving. This places a finite limit on the size of land animals (it is possible for animals that live in the sea, like the blue whale, to grow larger because the water supports some of their weight).

It is widely known that fleas and grasshoppers can jump to heights that are more than fifty times their body length. This is the equivalent of a man leaping 100 metres in a single bound. The world record for the human high-jump is much lower, a mere 2.45 metres, and if the jump is made from standing even a top athlete can only clear about 1.6 metres. So why can fleas and grasshoppers leap relatively so much higher? Their remarkable ability turns out to be simply a question of scale – it is physically impossible for a large animal to leap as high, in relative terms, as a small one. Indeed, physics predicts that similar types of animals should be able to jump to equal heights irrespective of their body size.

To understand why this is the case, recall that the muscles of a

Beyond the Limits

Fleas are renowned not only for the height but also for the rapidity which which they leap. The average acceleration a flea achieves during take-off is greater than 1350 metres per second per second – roughly the equivalent of a g-force of +200g. This is far faster than muscle can contract so how does the flea do it?

It turns out that the flea has a built-in catapult which it uses to store energy over a long time and then release it very quickly. Fleas have elastic rubbery material called resilin at the base of their hind legs. While the flea is at rest, muscle contraction gradually compresses the resilin, raising part of the hind leg into the air. The flea is now 'cocked', ready for take-off. When the trip mechanism is triggered, resilin expands rapidly and the powerful elastic recoil swings the leg down very fast and catapults the flea into the air.

The flight muscles of some insects also act beyond the limits. Each contraction of a mammalian muscle is initiated by a single nerve impulse. Insect flight muscles, however, contract far more frequently than nerve impulses can be conducted. The midges that make a warm summer evening in Scotland a misery beat their wings more than 1000 times a second, generating a high-pitched whine that humans can hear. This is more than forty times faster than human fast twitch muscles can contract.

Insect flight muscles make use of resonance to achieve these high rates of contraction. It turns out that their flight muscles are sensitive to stretch human and insect are able to exert the same force per cross-sectional area and that the cross-sectional area of the muscle determines its force. The mass (or volume) of an animal increases as the cube of its size, whereas the cross-sectional area of the muscle only increases as the square of its size. This means that, relative to its mass, the larger animal has less force available for jumping. A large animal could increase its jumping ability slightly if it increased the fraction of its mass that was made up by jumping muscle. This is in fact what the lesser galago, a small tropical primate, does. Relatively speaking, its muscle mass is twice that of a person. As a result, it is able to achieve a vertical leap from standing of 2.2 metres – roughly three times the height a man can jump (the human record for a standing jump is 1.6 metres but our centre of mass is about a metre above the ground at take-off). However, it is obvious that an animal can only devote a fraction of its body to muscle, so this adaptation is of limited use.

– if you pull on the muscle it contracts, and when you release the pull it relaxes. The thorax of the insect (the bit to which the wings attach) is a stiff box containing two types of flight muscle, one of which moves the wings up and the other which moves them down. Perhaps surprisingly, the flight muscles do not actually attach to the wings but instead are anchored to the walls of the thorax. Movement of the wings, which are attached to the roof of the thorax, is produced indirectly, by altering the shape of the thorax.

The thorax, in effect, acts as a resonating box that pulls alternately on the elevator and depressor muscles, stimulating first one and then the other to contract. When the elevator muscles contract, the roof of the thorax is depressed and clicks into a new position, causing the wings to move up. But the new shape of the thorax stretches the depressor muscles, causing them to contract, and it simultaneously removes the tension from elevator muscles, so that they relax. Consequently, the roof of the thorax suddenly snaps back to its original position, flicking the wings down. This, of course, stretches the elevator muscles once again, stimulating them to contract, and simultaneously relaxes the depressor muscles, so that the cycle begins all over again. Thus the roof of the thorax clicks backwards and forwards between two stable positions, moving the wings up and down as it does so.

Because movement of the thorax can be achieved with only tiny changes in muscle length, it can occur extremely fast. And because the flight muscles are stimulated by stretch, rather than nerve impulses, they can contract faster than nerve impulses can be conducted. This explains how insects are able to 'break the limits'.

Small animals also seem disproportionately strong. A dung beetle is dwarfed by the huge ball of manure that it pushes, and a leaf-cutter ant can carry a sail of leaf that weighs more than itself with ease. A man would find such a burden extremely onerous. The reason for the extraordinary power of ants is again a question of scale. The ant's muscles are only as strong as a man's but they appear much stronger because the force a muscle can exert relative to the body mass of an animal increases as an animal decreases in size. Relative strength is also simply a question of scale.

Taking the Strain

Regular physical activity, as we are constantly reminded, brings many benefits, among them a reduction in the risk of coronary heart disease, diabetes, obesity and osteoporosis. It makes us both look and feel better. But there is also a negative side.

Almost everyone who takes regular exercise, and many who do so only intermittently, suffers some form of overuse injury. Stories of shin splints, weak knees, strained muscles and stress fractures are commonplace. In weekend runners, it is usually a case of 'too much, too soon'. In elite athletes, it is 'too much, too long and too often'. Constant stress can fracture bones, most commonly those of the foot and lower leg, as is frequently seen in dancers and distance runners. Muscle strain leads to local inflammation, producing swelling and soreness. Friction injuries occur when tendons rub against the sheaths in which they are enclosed, or against the bones over which they pass, producing tendonitis in the knees and Achilles tendons. Repeated small tears in the tendon at the point of its insertion also lead to local inflammation. Tendons can sometimes tear completely, abruptly incapacitating the athlete. Torn ligaments around the joints can be particularly painful and debilitating – the knees are especially susceptible to this type of injury. Such overuse injuries require immediate rest and, following recovery, exercise must be reintroduced gradually and the training routine varied to avoid a recurrence. In the long term, the constant wear and tear produced by severe prolonged exercise may result in osteoarthritis, a chronic condition in which the joints degenerate, producing pain and stiffness. The human body is simply not designed to be used as a continuous running machine.

Stress also affects the immune system and professional athletes become more susceptible to infection, which compromises their performance. Female endurance runners and ballerinas may cease to menstruate, and the beneficial effect of exercise on their bones is then more than offset by the reduced oestrogen levels. This explains the paradoxical finding that young women who undertake strenuous exercise may develop osteoporosis, yet moderate exercise can slow bone loss in older women (see Chapter Seven). In young athletes, like gymnasts, exercise can also retard the onset of puberty.

Strenuous exercise can cause proteins to leak out of skeletal muscles, probably because of microscopic mechanical damage to the muscle cells themselves. This is quite normal. In a few cases, however, so much protein leaks out that it can be life-threatening. The victim feels nauseous, their muscles become swollen and ache, and their urine turns the colour of Coca-Cola because it contains myoglobin (the pigmented molecule, related to haemoglobin, that acts as a short-term oxygen store in muscle). Most dangerous of all is the fact that the concentration of salts in the blood becomes unbalanced. The condition is rare but occasionally seen in military recruits who perform multiple squat jumps as part of their initial training, hence its common name of 'squat-jump syndrome'.

Many sports also increase the risk of trauma. Bruised bodies and broken limbs are common in sports involving physical contact: rugby is notorious for broken noses, a hockey stick can easily break a leg, squash balls are just the right size to fit in the eye socket, and falling from a horse is a common cause of head injury. Even the spectator or passer-by is at risk. While cycling past a cricket field one summer afternoon, I was struck in the eye by a cricket ball and knocked off my bike. Next day, I had a magnificent black eye.

Bleeding in and around the point of impact causes pain and inflammation. This can be reduced by *ICE* – a mnemonic for *i*ce (which causes constriction of the blood vessels), *c*ompression and *e*levation (both of which reduce blood flow to the injured region). Recreational players who neglect this simple first-aid strategy by initially attempting to play on, and then nursing their injuries with a relaxing, but vasodilating, alcoholic drink should not be surprised that their sprained ankle is swollen, stiff and painful the next morning.

This litany of injuries is often invoked by the less enthusiastic as an excuse not to take any form of exercise. But it is wise to remember that while excess – as in so many other walks of life – may be deleterious, moderate exercise is highly beneficial. You may not succeed in being the fastest, or the strongest, but you will probably live an active life for longer.

6

THE FINAL FRONTIER

'I'll put a girdle round the earth in forty minutes.'

WILLIAM SHAKESPEARE, *A Midsummer Night's Dream*

Edwin 'Buzz' Aldrin standing on the surface of the moon on 20 July 1969. Neil Armstrong and the Apollo 11 lunar landing module *Eagle* can be seen reflected in his helmet visor.

THE DAWN OF 21 July 1969 is etched on my memory. Like millions of other people throughout the world, I sat gripped before a small flickering black-and-white TV screen, scored with a snowstorm of white lines and spots. We strained to hear the words though the hiss and crackle, but there was no mistaking the excitement and tension in the voices. Shivering in the dark unheated room, newly awoken from sleep, oblivious to how tightly I was clutching my cocoa cup, I was transported many thousands of miles away, captivated by a spine-tingling mixture of science, technology and exploration. I was seventeen years old and Neil Armstrong had just become the first man to set foot on the moon.

Step into the vacuum of space and you would perish in a few brief, agonizing moments. The air would rush out of your lungs; the dissolved gases in your blood and body fluids would vaporize, forcing apart your cells and forming bubbles in your capillaries, so that no oxygen would reach your brain; air trapped in internal organs would expand, rupturing your gut and eardrums; and the intense cold would cause instant freezing. You would be unconscious within less than fifteen seconds.

Man can only survive in space if he takes his environment with him but, even when protected by a spaceship, space flight presents several physiological problems. The first of these is the acceleration required to escape the Earth's gravity, which imposes an additional gravitational force on the body. The second is the opposite extreme – weightlessness.

This can cause disabling motion sickness, the redistribution of fluids around the body, a decrease in the number of red blood cells and a worrying loss of bone and muscle mass. If we are to realize our dream of travelling to other planets within our solar system, we must find a way to reduce these changes. In this chapter, we explore how space flight remoulds our bodies and how these changes can be ameliorated.

A Brief History of Space Flight

The space age began on 4 October 1957 when the Soviet Union launched the world's first satellite. They called it Sputnik, which, in Russian, means travelling companion. Within a month, Sputnik 2 followed, carrying a dog named Laika. And on 12 April 1961 the cosmonaut Yuri Gagarin sailed into the sky in Vostok 1, made one complete orbit of the Earth, was ejected from his spacecraft at an altitude of 7000 metres and landed safely by parachute. The whole trip took one hour and forty-eight minutes.

This impressive list of Soviet successes had a major impact in the United States. Although Eisenhower might off-handedly refer to Sputnik as just 'one small ball in the air', the general public (and the military) were less sanguine. Coming as it did at the height of the Cold War, they were shocked by the demonstrable superiority of Soviet technology. The continuous train of radio beeps emitted by the satellite as it flashed over the USA every ninety minutes served only to rub in this fact; it was, as Claire Booth Luce noted, a 'raspberry from Russia'. Almost overnight, the US government poured millions of dollars into science education and within nine months the country had a high-profile space programme of its own. The space race had begun in earnest. Yet it was not until 20 February 1962 that the first US astronaut, John Glenn, orbited the Earth. By this time another Soviet cosmonaut, Gherman Titov, had followed Gagarin's lead and looped around the Earth an impressive seventeen times; and a year later Valentina Tereshkova became the first woman in space.

The Americans were not to be outdone and rapidly raised the stakes. The Apollo space programme was initiated by President John Kennedy's challenge that the United States 'should commit itself, before this decade is out, to landing a man on the moon and returning him safely to Earth'. Broadcast in 1961, his speech meant that this goal

Yuri Gagarin (1934–68), the first man in space, in the cabin of his spacecraft Vostok 1

had to be accomplished within a brief nine years. The speed at which the necessary technology was developed was striking. Christmas 1968 found Frank Borman, Jim Lovell and Bill Anders in lunar orbit and, less than a year later and well within Kennedy's deadline, came the first lunar landing. Yet a mere three years on, after only six lunar explorations, the moon was abandoned – not for scientific reasons but for political ones. Today, it sometimes seems unbelievable that men ever walked on the moon and that for a few brief hours the world was held spellbound by their achievement.

Rather than aiming directly for the moon, the Soviet strategy was to build an orbital space station that could be used as a staging post for flights outside the Earth's gravitational field, and in which cosmonauts could live and work for extended periods. The world's first space station, Salyut 1, was launched by the Soviet Union in 1971, and remained in orbit for just over two years. It was followed by further Salyut craft and then, on 20 February 1986, by the space station *Mir* (which means both 'peace' and 'world' in Russian). Designed for a life of five years, *Mir* surpassed expectations and remains in orbit to this day, albeit in a precarious condition and plagued by continual faults. After the disintegration of the Soviet Union in 1994, Russian cosmonauts and US astronauts conducted many joint missions aboard *Mir,* but it is now uninhabited and will soon be allowed to burn up in the Earth's atmosphere. It is to be replaced by an International Space Station, built by a consortium of many nations.

The different policies adopted by the USA and Soviet Union meant that studies of the long-term effects of life in space were, until very recently, largely confined to the Soviets, and that the record for the longest flight is held by a cosmonaut, Valerie Polyakov, who spent 438 days on the *Mir* space station between 8 January 1994 and 22 March 1995. Both countries, however, have amassed a considerable amount of information about the short-term effects of space flight.

Eyeballs In and Eyeballs Out

The first problem that confronts an astronaut is the acceleration experienced during the launch, as the spacecraft is rocketed from rest to orbital velocity.[1] By itself, speed itself has no obvious effect on the human body. Even as you sit quietly reading this book, you are travelling at 108,000 kilometres per hour through space and spinning at a rate of up to 1670 kilometres per hour,[2] as the Earth orbits the sun and turns on its axis. Sealed inside an aircraft, in the absence of visual clues, it is also difficult to appreciate that you are moving at high velocity if the plane is travelling at a constant speed in a straight line. It is a very different matter, however, if the plane dives or banks steeply. This illustrates the fact that our bodies are designed to detect changes in speed or direction and accommodate quickly if nothing alters.

Acceleration is measured in terms of g-force, where +1g is the pull of the Earth's gravity on the surface of the planet. Linear acceleration is defined as a change in speed without a change in direction, while radial acceleration is a change in direction without a change in speed. Most people are familiar with what linear acceleration feels like; this is the force that thrusts you back in your seat during a racing start in a sports car, or when an aircraft takes off. Much greater g-forces are experienced during the catapult launch of a plane from an aircraft carrier, during space launches, or when a car crashes into a brick wall at speed. Radial acceleration is produced when motorcyclists race round the Wall of Death or when an aircraft banks steeply. A change in direction in a commercial airliner typically results in about +1.3g, but accelerations of up to +8g can be reached in high-performance military aircraft during steep turns. Usually, aircraft make head-to-centre turns, which cause the blood and internal organs to fall towards the feet. This is known as a positive g-force because it operates in the same direction as the Earth's gravity. Occasionally, aircraft make head-out turns that force the organs and body fluids towards the head. This is known as negative g, or, more colloquially, as the 'eyeballs out' position (positive g being known as 'eyeballs in'). You can experience -1g for yourself quite easily, simply by standing on your head. The nearest most people come to higher g-forces is at a funfair, where some state-of-the-art rides can generate g-forces as high as +4g. This is what clamps you in your seat when the cart turns upside-down during the giant loop-the-loop, that flypapers you to the wall in a spinning centrifuge, and that (in its negative form) leaves your stomach in your mouth as a rollercoaster dives precipitously downwards.

The question of how much g-force the human body can withstand is of considerable interest to the world's air forces, for the power and agility of military aircraft are now limited by the physical abilities of the pilot. The usual way to investigate the effects of increased gravitational force on humans is to spin them in a centrifuge. This machine operates on the same principle as a spin-dryer, where the centrifugal force caused by rapidly spinning the drum flings the clothes outwards to the edge of the drum and forces the water out of them. In a human centrifuge, the person is strapped in place to prevent them from flying off, but their body fluids may shift in response to the increased g-force. The subject sits in a pivoted cabin that swings out when the centrifuge is spun, so that their head ultimately points towards the centre of the machine and they experience a positive g-force that tends to drag the blood down towards the feet. Would-be fighter pilots and astronauts are tested in such centrifuges for their ability to withstand high g-forces.

As the g-force is increased, you become less and less capable of useful function. At +2g, the body feels heavier, the facial tissues sag, and it is difficult to stand up from sitting. At +3g, standing becomes impossible, and as the g-force increases further a grey curtain gradually sweeps inwards from either side of the head as colour vision fades out, beginning at the periphery of the eye. Sight is lost completely around +4.5g, although it is still possible to hear and think. At +8g, it is impossible to raise your arms or lift your head. Somewhere around +12g most people lose consciousness, slumping down in their seats, their heads lolling on their shoulders. Convulsions may occur during this period, or on deceleration, which are colloquially known as 'doing the funky chicken'. Currently, US air-force recruits must successfully withstand +7.5g for sixteen seconds in order to become a fighter pilot. However, even if they do not black out, an aviator exposed to such g-forces would be unable to escape from an aircraft unaided, so an ejector seat is essential.

Our bodies are well adapted to Earth gravity. Most of the time we are unaware of its presence, although we are increasingly reminded of its action as we age, by the inevitable sagging of skin and flesh and the development of varicose veins. Higher g-forces are quite another matter. A positive g-force may pull the blood towards the legs so strongly that the heart is unable to pump against it effectively, leading to a reduction in the blood supply to the brain and loss of consciousness. Breathing difficulties are also encountered because the diaphragm is pulled down, making expiration more difficult. Consequently, ventilation of the lower part of the lungs is reduced. The problem is compounded by the fact that there is less blood available to perfuse the upper part of the lungs because of the effect of the increased g-force on the circulation. Positive g-forces therefore cause a large decrease in gas exchange at the apex and base of the lungs.

To overcome these problems, military pilots are trained to carry out breathing and straining exercises. They tense the muscles in their legs, which squeezes the leg veins and helps force blood back to the heart and brain. When flying high-performance jets such as the Tornado and F16, it is naturally more difficult to perform such exercises so pilots wear anti-gravity trousers that do the job for them. These inflate at high g, providing an external pressure that squeezes the legs and helps return the blood to the heart. Centrifuge tests have shown that, as might perhaps have been expected, shorter people are able to tolerate the highest positive g-forces. Taller people are at a physiological disadvantage as they have a greater distance between their heart and their brain.

Negative g-force is encountered less frequently, but it is also unpleasant. It drags the blood to the head, causing the small blood vessels to bulge and burst under the strain, producing a condition known as 'red-out' that sometimes afflicts bungee-jumpers.

The Human Yo-yo

Bungee-jumping originated in Britain with a group of daredevil Oxford undergraduates who formed the Dangerous Sports Club. The first jumper was Bing Boston, an American student who leapt off the Clifton suspension bridge attached to a long piece of elastic in April 1979. To mark the importance of the occasion, he was appropriately dressed in white tie and tails. The Dangerous Sports Club got the idea from the manhood initiation rituals of Vanuatan South Sea islanders who leapt from the top of 35-metre-high rickety wooden towers with vines tied to their ankles, being brought to a jarring halt with their heads just inches above the ground.

One of the best known of all bungee-jumps must be that in the film *Goldeneye*, where James Bond (*aka* the stuntman Wayne Michaels) launched himself from the edge of the Verzasca dam in a perfect swallow dive. It took him less than six seconds to drop 183 metres (600 feet). Arguably the most famous of all bungee-jumpers, however, is the New Zealander A.J. Hackett, who jumped off the Eiffel Tower in June 1987, dived 300 metres from a helicopter in 1990, and in October 1998 leapt from the top of a skyscraper in Auckland. New Zealand is a favoured spot for bungee-jumpers, and many tourists test their nerve from the bridge that soars 80 metres over the gorge of the Rangitikei river.

Gravity will cause a bungee-jumper, or a sky-diver, to accelerate as they fall. The maximum rate of acceleration due to gravity is 9.8 metres per second per second and this is the rate at which you drop towards the ground in free fall. The problem for bungee-jumpers is not the rate at which they fall but the deceleration that occurs when the snaking bungee straightens and snaps taut as it reaches the limit of the elastic. The g-force can be quite severe and causes the blood to rush to the head, which may precipitate haemorrhages in the eye or even lead to detachment of the retina. These problems do not arise for sky-divers because the deceleration when they open their parachute is not so abrupt and, more importantly, the head is uppermost.

Taking Off

The g-force experienced by an astronaut varies throughout the launch, because it is governed by Newton's Law of Motion which states that force = mass × acceleration. Lift-off tends to be fairly gentle because the thrust of the rocket only just exceeds the weight of the spacecraft. The greatest g-forces occur as the spacecraft enters orbit, when it is much lighter (because most of the fuel has been spent) and the rockets are still at full thrust.

Early astronauts endured considerable g-forces. During the launch of the Mercury-Friendship 7 flight in 1962, for example, John Glenn was subjected to more than +6g for ninety seconds, and to a brief period where acceleration climbed as high as +8g. Glenn lay on his back with respect to the Earth, so that the gravitational force was imposed in the chest-to-back direction, in order to avoid the dramatic effects sustained when the g-force is imposed from head to toe. Even so, as one astronaut commented, 'it feels as if an elephant were sitting on your chest'. Among the greatest g-force ever experienced by cosmonauts was that encountered during the launch of a Soyuz spacecraft in September 1983. Because a fire broke out below the rocket ninety seconds before take-off, the launch had to be aborted and the emergency escape system shot the capsule about a kilometre into the air, subjecting the crew to g-forces as high as +17g. They survived the ordeal unscathed, landing safely by parachute some distance away. The g-forces encountered by present-day astronauts are far more gentle. The crew of the Space Shuttle, or the Soyuz craft that supply the *Mir* space station, normally never experience more than 3.5 times the Earth's gravitational pull during launch.

Military pilots endure even greater g-forces than astronauts (a massive +25g) if they have to make an emergency ejection from a crippled aircraft, although these are sustained for a far shorter time. Pulling the firing handle first jettisons the plane's canopy and then fires an ejection gun mounted beneath the seat, launching the pilot, still strapped in his seat, into the air. Clearly, the faster the pilot can be shot out the better, but too high an acceleration damages the human spine. As a result of experiment and practical experience, it has been determined that the peak acceleration should not exceed +25g – if it does, the risk of spinal injury increases dramatically. The most modern ejector seats incorporate rockets that continue to burn for around half a second after launch, which enables the peak g-force, and thus the risk of back injury, to be reduced.

An additional problem encountered during space launches is severe vibration. Shaking and jarring of the body is not only uncomfortable, it can impair the ability to perform manual tasks, precipitate nausea and may cause the body to resonate in time with the oscillations. For poorly understood reasons, this can lead to hyperventilation and physical collapse.

Life Support

A spacecraft must protect its crew from the extremes of space. Seven hundred kilometres above the Earth's surface, the number of gas molecules is infinitesimally small and the pressure approaches that of a perfect vacuum; a spaceship must therefore provide both a breathable atmosphere and protection against the extremes of pressure. Space is also extremely cold, almost -270°C, but solar rays heat up objects in their path fast, so spacecraft must have a temperature control system that is able to cope with extremes of heat and cold. Damage from micrometeoroids or space debris is a constant concern. Even a small flake of paint chipped off a satellite and travelling at several thousands of miles an hour can fatally puncture a spacecraft. The windows of the Space Shuttle are so regularly dented by micrometeoroids that they must be renewed every few flights.

In 1998, a supply ship hit the space station *Mir*, creating a tiny hole smaller than a postage stamp. The air screamed out into the void, but fortunately the hole was small and the rate of loss sufficiently slow that the astronauts were able to seal off the leaking compartment. The crew of Soyuz 11 were not so lucky. On returning to Earth, their descent capsule made a perfect automatic landing, so the recovery crew were horrified to find, on opening the capsule, that the cosmonauts were dead. It transpired later that a pressure equalization valve had accidentally opened in orbit, shortly after the descent capsule had separated from the orbital module. Because the crew had removed their pressurized spacesuits in order to squeeze into the tiny descent capsule, they were killed by asphyxiation. Today, astronauts wear protective suits during take-off and landing to guard against a possible loss of pressure but while in orbit they wear ordinary clothing, which enables them to move about more easily.

The crew of early US spaceships breathed pure oxygen, at a pressure of one-third of an atmosphere. This strategy enabled them to carry more oxygen, weight for volume, than if they used air with the same composition as the Earth's atmosphere (which contains 78 per cent nitrogen). Although pure oxygen is toxic if breathed for more than twenty-four hours at atmospheric pressure (see Chapter Two), it is quite safe at one-third of an atmosphere. In the Mercury and Gemini missions, the spacecraft was filled with pure oxygen at a pressure of one atmosphere while on the launch pad, and the pressure was reduced on reaching Earth orbit. This practice was subsequently altered as a consequence of a disastrous fire that occurred during a routine simulated launch of Apollo 1 and killed astronauts Gus Grissom, Ed White and Roger Chafee. At atmospheric pressure, pure oxygen is a considerable fire hazard. What seems to have happened to Apollo 1 is that a stray spark ignited flammable material within the cockpit, and the oxygen atmosphere quickly turned the command module into an inferno. Following this tragedy, all Apollo spacecraft used an Earth-normal atmosphere during launch and switched to pure oxygen only after reaching orbit. In contrast, Soviet spaceships have always been pressurized to one atmosphere and used air with a similar composition to that on Earth: 78 per cent nitrogen and 21 per cent oxygen. This strategy has now been adopted by NASA also, partly because of worries about the effect of breathing pure oxygen, even at reduced pressure, for extended periods on long duration flights.

Respiration increases the concentration of carbon dioxide in the air, which can lead to headaches, drowsiness and eventual suffocation (see Chapter Two). It must therefore be removed. In spacecraft, this is achieved by chemical reaction with lithium hydroxide (which is converted to lithium carbonate in the process). Lithium hydroxide canisters and the dangers of carbon dioxide accumulation shot into the limelight in April 1970. Two and a half days into the Apollo 13 mission, disaster struck. An electrical short caused one of the three fuel cells that powered the command module to explode. In turn, the force of the explosion severed the supply from the other two fuel cells, depriving the spacecraft of all power. The lunar landing craft *Aquarius* became the astronauts' life-raft, supplying oxygen, water and electrical power. Unfortunately, it was only equipped with sufficient lithium hydroxide canisters to cleanse the air of carbon dioxide for two men for two days, while it was going to take more than three days to get back to Earth and there were three crew members. News bulletins around the world quickly introduced the public to the perils of excessive carbon dioxide. In fact, plenty of lithium hydroxide canisters were present in the command module but they could not be used by the *Aquarius'* air-purification system because they were the wrong shape. Working round the clock, a team of engineers on Earth figured out a way to rig up a makeshift air-purifier using the wrong-shaped canisters and an eclectic mix of cardboard, plastic bags, sticky tape and old socks. As a child growing up in Britain, my favourite TV programme was *Blue Peter*, which showed you how to make models from yoghurt pots and elastic bands. The makeshift air-purifier of the Apollo 13 mission was the ultimate *Blue Peter* creation. Fortunately, it worked.

Breathing also generates water vapour, as is clear to anyone who has sat in a car with the windows closed in cold weather; the mist that collects on the inside of the windows is largely due to water expired from your lungs. The amount of water vapour in the air of a spacecraft must be carefully controlled, for too much causes condensation, while too little leads to drying of the eyes and mucous membranes in the throat. To achieve a satisfactory atmosphere, the air in the spacecraft is constantly recirculated, carbon dioxide and dust particles are filtered out, and the humidity and oxygen concentration adjusted as necessary.

Inside a spacecraft, the temperature is kept at a comfortable 18–27 °C. Temperature control is critical because the spacecraft is cooked by the sun on one side and frozen by the cold of space on the other. When the *Mir* space station lost all power, it became icy cold when the sun vanished behind the Earth, but unbearably hot once it reappeared. To help maintain a constant temperature during the journey from the Earth to the moon and back, the Apollo spacecrafts were slowly rotated, a motion that not surprisingly came to be known as the 'barbecue roll'. In the Space Shuttle, heat loss is achieved by 'space radiators' located on the insides of the cargo bay doors, which are opened once the Shuttle reaches orbit.

Free Fall

Although we take most of our environment with us into space, gravity is another matter. There is little incentive to develop artificial gravity, partly because one goal of space research is to get away from Earth's gravity and partly because, for short flights at least, the effects of microgravity are not debilitating. Nevertheless, the physiological stress of weightlessness is not negligible. It causes an immediate shift in body fluids from the legs to the chest and head, and it impairs the balance system, which can precipitate space sickness. With prolonged flights there is also a progressive loss of red blood cells, leaching of calcium from the bones and muscle degeneration. Most of these changes stabilize within about six weeks, but bone loss continues throughout the flight and no adaptation has been observed even in flights as long as a year.

The pull of gravity in an orbiting spacecraft is actually not very different from that at the Earth's surface. The reason that the crew experience weightlessness is because they are in continuous free fall. On Earth, we feel weight only because the ground below us pushes up to prevent us accelerating towards the centre of the Earth. Whenever this reaction is removed – when a skydiver leaps from a plane or for the split-second that we are in mid-air when we jump off a wall – we experience weightlessness. In effect, an orbiting spacecraft is in constant free fall but as it falls towards the Earth its speed carries it on past, so that it continues to orbit. For accuracy, therefore, conditions during orbital flight are referred to as microgravity rather than zero gravity.

The lowest orbits are 200 kilometres above the Earth's surface, the altitude at which air resistance becomes insignificant. At lower altitudes the drag of the Earth's atmosphere slows down the spacecraft so much that it eventually spirals downwards and burns up in the lower atmosphere. The space station *Mir* orbits at around 400 kilometres above the Earth's surface, but even at this height, its orbit constantly decays and corrections have to be made every few weeks to regain the original level. The upper limit for manned Earth orbits is set by the need to avoid the ionizing radiation belts which encircle the planet about 400 kilometres above its surface (see later).

Weightlessness

Weightlessness has a marked effect on the distribution of body fluids. On Earth, gravity causes the blood and tissue fluids to pool in the legs and lower body, but as soon as you escape from the Earth's gravitational field these fluids shift upwards, producing some very obvious and uncomfortable changes. Your face puffs up, your neck and facial veins stand out prominently, your eyes feel as if they are bulging out, your nose becomes blocked and your sense of smell and taste diminish. The general feeling is somewhat similar to that encountered when 'full of a cold'. Another consequence is that your legs shrink, losing about a tenth of their volume – decreases in the circumference of the calf of up to 30 per cent have been reported. Cosmonauts sometimes wear elastic straps around the tops of their legs to restrict the flow of body fluids from the legs (because the blood pressure in the arteries is greater than that in the veins, the flow of blood into the legs is not impeded).

Pressure sensors in the head and chest are stimulated by the fluid shift, and within a few days the body readjusts to the effects of weightlessness by reducing the volume of blood and body fluids through increased urination and decreased fluid intake. Astronauts find they lose weight during the first few days in space, largely because of this loss of body water. The urge to urinate more may be very inconvenient, especially if, as on the early space flights, the astronaut is wearing a spacesuit. There is no evidence that the shift in body fluids towards the

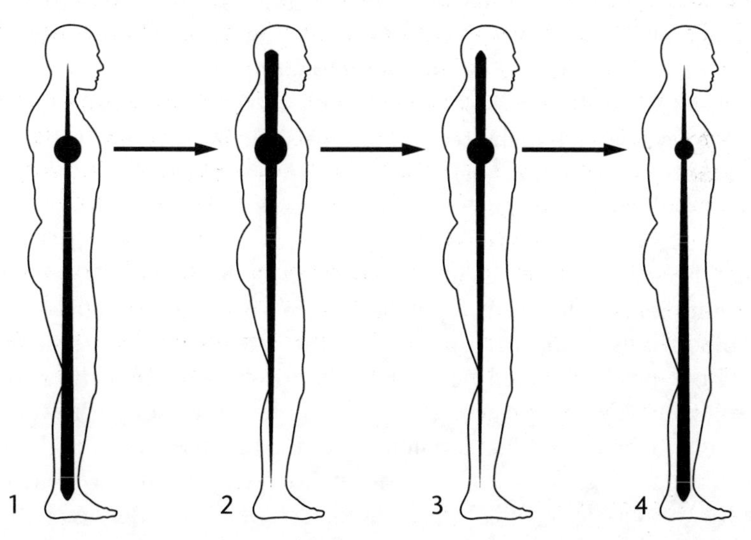

The shift in body fluids during weightlessness. On Earth, fluid accumulates in the lower half of the body because of gravity (1). Within a few minutes of arriving in a weightless environment, about 2 litres of body fluids migrate to the chest and head (2). Compensatory mechanisms then lead to a gradual redistribution of fluid throughout the body (3). On return to Earth, gravity once more exerts its influence but because of the readjustments that take place in space, comparatively more blood pools in the legs than normal (4). This may make it difficult for the returning astronaut to stand up without blacking out.

head, or the compensatory mechanisms that result as a consequence, impair cardiovascular function in space. It is a very different matter on return to Earth, however, as discussed later.

Liberated from the stress of gravity, astronauts grow taller, because the cartilaginous disks that separate the bones of the spine are no longer compressed. Most people manage one or two centimetres but some individuals may grow much more, as was the case for John Glenn on his second space flight, at the age of seventy-seven, who found he had added an extra 6 centimetres (a couple of inches) to his height. Engineers must take these changes in size into account. On one Shuttle flight to investigate the effects of weightlessness on the nervous system, the designers of an experimental chair used to measure the astronauts' reactions forgot to allow for expansion and the astronauts complained that the chair became too small. The lungs, heart, liver and other organs, also no longer weigh anything and float about in the body cavity. As one astronaut memorably put it, 'You feel your guts floating up.'

Red blood cell production is markedly decreased in microgravity. Red blood cells have a brief life – only 120 days – so a reduction in their manufacture leads to a drop in the number of red blood cells in the circulation. This decline starts within four days of exposure to weightlessness and stabilizes around forty to sixty days later. During one ten-day Spacelab flight, red cell numbers fell by around 10 per cent, and larger decreases have been observed on longer flights.

As explained in Chapter One, the production of red blood cells is controlled by the hormone erythropoietin, whose secretion is determined by the oxygen level in the tissues, and the higher the oxygen concentration, the less erythropoietin is released and thus the fewer the red blood cells produced. Initially, it was therefore thought that red blood cell production fell because the atmosphere of the early space capsules was high in oxygen. This idea had to be reassessed however, when it was discovered that the number of red blood cells still declined on later space flights, where Earth-normal oxygen pressures were used. The loss of red blood cells is now thought to be a consequence of the changes in blood volume produced by microgravity. It is believed that the shift of blood into the chest that occurs on exposure to weightlessness tricks the body into thinking that it has too much blood and causes it to reduce the production of red blood cells. This is mediated by a marked drop in erythropoietin levels. Simply decreasing production, however, is not sufficient to account for the striking loss of red cell mass on exposure to microgravity; in addition, red blood cells waiting to be released from the bone marrow are actually destroyed.

Sleep

Astronauts often complain that they find it difficult to sleep in space. Undoubtedly, some of this difficulty must be due to the novelty of space flight. Furthermore, spacecraft can be noisy and one's colleagues left on watch may not be silent. But it seems likely that the main cause of sleeplessness is disruption of the body's normal circadian rhythm (its biological clock). Many physiological processes, among them sleep, are controlled by circadian rhythms which, in turn, are regulated by the light–dark cycle. It is well established that people in northern latitudes sleep far less during the Arctic summer, when the sun never sets, than they do during the continual darkness of the Arctic winter. Because the sun rises and sets once every ninety minutes as the Space Shuttle orbits the Earth, the astronaut's normal light–dark cycle is markedly affected.

Sleeping in microgravity also poses other problems. To ensure that they do not drift around the spacecraft while sleeping, astronauts must zip themselves into sleeping bags attached to its walls. Most people sleep best when they feel secure, but in microgravity there is little sense of pressure so you do not feel as if you are lying in bed. Some astronauts find it easier to sleep if they put a strap across their forehead, so that it feels as if their head is resting on a pillow, and they may use similar straps across their knees to help them curl up. They must also be careful to position themselves in a continuous flow of air, so that the carbon dioxide they exhale blows away and does not accumulate around their face and suffocate them. On Earth, breezes and convection currents ensure a continuous circulation of fresh air, but in microgravity there is no convection to carry away the exhaled carbon dioxide, for warm air does not rise (warm air is no longer lighter than cold air in space, since neither has any weight).

Infection

Each of us is host to millions of microorganisms that accompany us wherever we go, even into space. It is estimated that a healthy individual

Living in Microgravity

Most people delight in microgravity, describing it as the ultimate freedom. It is possible to float under the table, stretch out on the ceiling (although the terms 'floor' or ceiling' cease to mean anything), hang suspended at the centre of the turning world or fly gracefully around the cabin. Acrobatic manoeuvres, such as somersaults and spins, become easy, even for those with no gymnastic training. Because it is possible to move in three dimensions, the cramped capsule suddenly seems spacious.

Movement in microgravity, however, is not straightforward. To move, you must push off the walls of the cabin, rather like a swimmer uses the wall to turn at the end of the pool, but if you push too hard, you will move too fast and cannon into the opposite wall. Novice astronauts get many bruises before they learn to use their fingertips to push themselves off gently.

Liberated from gravity, objects thrown in space move in a straight line, not the falling arc they follow on Earth. In her autobiography, Helen Sharman described how she took her first drink in space, not by using the special mouthpiece provided, but by catching in her mouth a shimmering, wobbling bubble of water, fired from a pressurized water tank by her grinning colleague: 'I closed my mouth around it and was rewarded with a delicious explosion of cold water.'

Microgravity beautifully illustrates the difference between mass and weight. Mass is the resistance of an object to movement, while weight is the effect of gravity on mass. In space, weight vanishes but mass remains. This is why you can balance a man or a mouse on your little finger equally easily, but if you try to push them from one side of the cabin to the other, you will find the mouse requires less effort to move.

Sir Isaac Newton's Third Law of Motion states that 'for every action there is an equal but opposite reaction'. This is not always evident on Earth – when we lift an object, or push one way from us, we remain static because the planet on which we stand is massive and resists movement. The situation is very different in space. If an astronaut pushes an object of similar size to himself, both of them move – in opposite directions. If he tries to turn a nut with a spanner, the nut remains firmly fixed in place, while the astronaut spins around it. An astronaut must therefore have his feet firmly planted on a stable surface that does not move. Foot restraints are used to anchor him in place, and are vital for work outside the spacecraft to prevent the astronaut from drifting away from the spaceship and on out into space.

Some activities are particularly

difficult in microgravity. Washing is a problem because the water droplets fill the air, forming glistening, quivering spheres that float in the cabin and are hard to remove. They slip through the fingers and dissolve into myriads of smaller globules. Astronauts have to make do with sponge baths.

Water may be fun to play with, but other liquids are less pleasant to clear up. One of the great challenges for spacecraft engineers has been to design effective and acceptable space toilets. Early missions relied on in-suit collection devices but these have been superseded by space toilets that function in very much the same way as ones on Earth, except that they use suction to pull the urine droplets in. These are then released into space where they instantly freeze to form a cloud of glittering ice crystals. When asked what he found the most beautiful sight in space, one of the Apollo astronauts answered 'urine dump at sunset'.

Solid waste must also be removed by vacuum, and is then stored and returned to Earth for final disposal. Shaving, even with an electric razor, fills the air with fine hairs so shaving cream (to stick them together) or a vacuum cleaner are an essential accompaniment. While it is no longer necessary to set your camera down on a table, because it will float happily in mid-air beside you, anything left unattached drifts away at the slightest touch and therefore must be anchored down with Velcro or elastic straps.

Housekeeping is a nightmare in space, for dust never falls but lingers in the air. The spacestation *Mir* is well ventilated and the circulating air is filtered, but it is still littered with minute dust particles formed from flakes of skin, stray hairs, and microscopic food particles. Humans shed about 10 billion skin scales every day. On Earth, these contribute to the white dust that collects on exposed surfaces in your bathroom, but in space they remain floating in the air you breathe. Consequently, cosmonauts tend to sneeze a lot – sometimes as much as thirty times an hour. Eye irritation from internal air pollution is also a common complaint.

More exotic, perhaps, is the fine black dust, like soot, that covers the surface of the moon. This was a major difficulty for the Apollo astronauts because, inevitably, they carried it into the landing craft on their boots. On the moon, where gravity is one-sixth that of the Earth, it fell slowly to the ground unnoticed, but once in space it worked its way into everything and turned their spacesuits black. Strangely, it smelt of gunpowder. Moon dust was not just an aesthetic problem; the fine particles could clog zippers on spacesuits, cause switches to stick, prevent electronics from working correctly, and coat the inner surface of the astronauts' lungs. Another concern was that it might contain microbes that could contaminate the Earth.

has more than a thousand billion (10^{12}) bacteria on their skin and many millions more in their gut. As many as 10 million of them are shed along with skin scales every day. The aphorism 'coughs and sneezes spread diseases' is even more apt in space. On Earth, droplets crowded with bacteria settle quickly to the ground where they do little harm, but in the absence of gravity they float in the air forming a fine aerosol that can be breathed in by other astronauts. Early space missions were plagued by minor illnesses – over 50 per cent of the crew suffered from skin, gut or respiratory infections – but isolation of the crew on the first Apollo flights prior to flight and scrupulous disinfection of the spacecraft both before and during flight have markedly reduced the incidence of infection.

Space Sickness

When astronauts first enter space, their movements are uncoordinated and they find it difficult to grasp objects accurately (they tend to 'overshoot' their targets). Many also report sensations of tumbling or rotating – as if they are turning upside-down – and they may suffer from vertigo. More seriously, around two-thirds of astronauts experience space sickness, some of them quite severely. The symptoms include headaches, nausea, dizziness, loss of appetite, lack of motivation, drowsiness and irritability. Vomiting can occur very suddenly, often without prior warning, and usually in infrequent bouts between which the victim feels almost normal. Space sickness can be a very real handicap for astronauts, preventing them from carrying out their work, and it is potentially lethal when wearing a spacesuit. Of special concern is the fact that most susceptible individuals succumb within the first hour of exposure to microgravity, during the critical early stages of the mission. Fortunately, most astronauts recover after two or three days in space.

Space sickness is most commonly precipitated by tilting the head forwards or backwards, although in some cases a visually disorientating scene may be sufficient. If you suffer from sea sickness, you will have already discovered that you fare best on deck where you can fix your eyes on the horizon. Astronauts have a far more difficult problem, because all visual reference points are arbitrary. There is no 'up' or 'down' in space. It is a topsy-turvy world, where the reference point constantly switches, in the same way that it does when looking at the famous Wittgenstein duck–rabbit paradox. While some astronauts find this disorientating initially, others adapt quickly. John Glenn commented: 'Before the flight, some doctors predicted that I might have uncontrollable nausea or vertigo, when the fluids in my inner ear were free to randomly move about during weightlessness . . . But I had no such problems . . . I found weightlessness to be very pleasant.' But Glenn was strapped down in his seat throughout his brief flight. In contrast, present-day astronauts are free to move about and the less fortunate experience space sickness when they see a crewmate floating 'upside-down' or when experimenting with acrobatic manoeuvres.

Although the exact cause of space sickness is unknown, it is believed it may result from conflicting signals about the position of the body. Our sense of orientation is derived from integrating information from balance organs in the inner ears with that from muscle and joint receptors, that signal the position of the limbs, together with visual cues. In space, many of these receptors do not receive their normal stimuli. Visual cues, in particular, lose their usual meaning. The Space Shuttle, for example, flies upside-down in relation to the Earth, with its tail fin pointing towards the planet. The crew try to maintain the normal Earth orientation in the cabin for the first few days in space (i.e, they actually fly upside-down), to help them adjust to the disorientating effects of weightlessness, but later, as they become more accustomed to their new environment, they orientate themselves at random.

Paying the Price

The long-term consequences of microgravity include bone loss and muscle wasting, both of which may be very substantial on long-duration space flights. While this does not noticeably impair performance in flight, it may have serious consequences on return to Earth. It can take a long time to restore bone and muscle mass to their pre-flight levels – roughly as long as the space flight itself – and it is not known whether they will recover completely after very long space flights, such as those required for the journey to Mars.

Bone is a living tissue that is constantly being remodelled throughout our lives. The more stress placed on bone, the thicker it becomes and, conversely, when the load is reduced – as when one escapes the

Balance Points

We have two balance organs, one on each side of the head. They are known as the vestibular apparatus and are found in the inner ear. Each comprises two otolith organs and three semicircular canals, and provides information about the movement and position of the body.

The otolith organs are fluid-filled sacs with sense organs embedded in their walls. These consist of groups of cells that have numerous fine sensory hairs, known as cilia, on their upper surfaces. The cilia project into a layer of gelatinous material covering the surface of the cell. On top of this lie tiny crystals of calcium carbonate, known as otoliths (literally, ear stones), that are the size of grains of dust. They act as gravity detectors.

When the head is held erect, the cilia stand upright, supported by the layer of jelly that covers the cell surface. If the head is tilted, however, gravity causes the crystals to slide sideways, pulling on the sensory hairs and stimulating them. The otoliths are also sensitive to vertical forces: they lift off the sensory hairs when you drop downwards in a lift and provoke the sensation that your stomach has been left behind.

In space, gravity no longer compels the otoliths to rest on the sensory cells, so that the brain receives conflicting information about the position of the body in space from the otolith organs and the eyes. It is believed that this conflict contributes to space sickness.

Angular acceleration is detected by the semicircular canals. There are three of them, orientated at right angles to each other along the X, Y and Z axes, which enables us to detect movement in three different planes corresponding to nodding the head (pitch), tilting it to one side (roll), and shaking it from side to side (yaw).

Each semicircular canal is a hollow fluid-filled tube with a swelling (the ampulla) at one end, within which sensory cells are located. The sensory cells bear many fine cilia on their surface, which project into the centre of the canal. If the fluid that fills the canal moves relative to the canal wall, the cilia are distorted, stimulating the sensory cells.

When you start to turn your head, your skull moves but the fluid in the semicircular canals gets left behind because of inertia. Consequently, the cilia will be bent by fluid drag, stimulating the sensory cells and producing the sensation of motion. If you continue to turn, the fluid will eventually catch up and move at the same rate as the skull; at this point, you will no longer experience the feeling of turning. This means that the semicircular canals detect changes in angular velocity, but are insensitive to sustained rotation. A pilot in a turning aircraft, for example, ceases to be aware that they are turning within fifteen to thirty seconds and must rely on instruments and visual clues to appreciate their situation.

When you stop rotating, the fluid

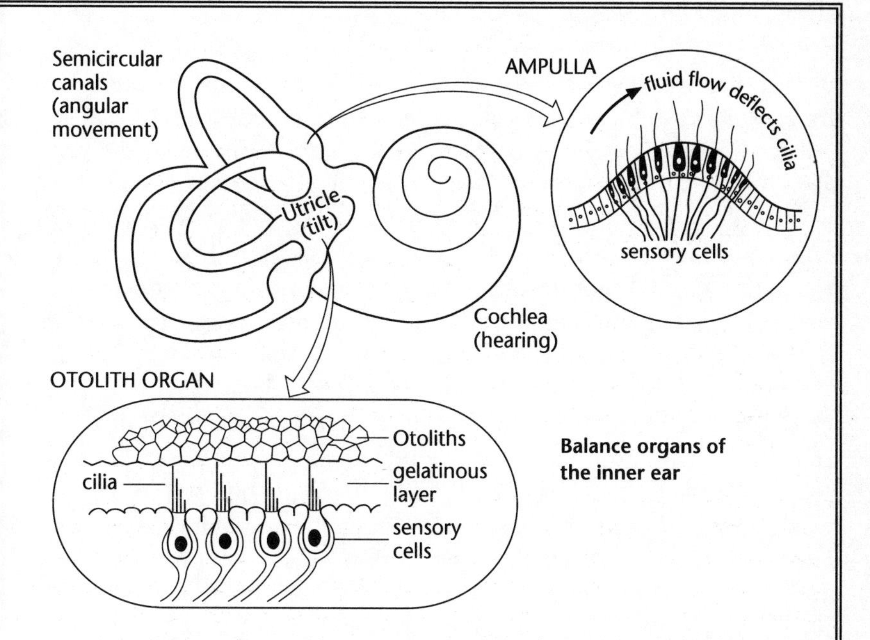

Balance organs of the inner ear

in your semicircular canals has built up momentum and continues to rotate for a while, leaving you with the sensation that you are still moving. This is why a pilot coming out of a spin has the overwhelming feeling that they are turning in the opposite direction. A similar effect can be observed if you circle on the spot for several seconds and then stop.

On Earth, when the head is tipped up and down, or tilted to one side, both the gravity and angular acceleration detectors are stimulated. In microgravity, the gravity receptors no longer respond but angular acceleration is still detected. The brain therefore receives a different signal from that expected, which may account for why space sickness is often triggered by movements of the head. With time, the brain accommodates to the conflicting signals and space sickness disappears.

Information from the vestibular system is coordinated with eye movements to ensure that the world appears to remain stable when the head is turned. When you turn your head to the right, a compensatory reflex rotates your eyes to the left at the same rate, so that you see a constant image of the world. The link between the vestibular apparatus and eye movement is the reason the world appears to spin when you stop rotating: rather than the world turning, your gaze is moving in the opposite direction.

Astronauts report that when they move their heads during space flight, it seems as if the world moves rather than themselves, suggesting that the link between the vestibular apparatus and eye movement is affected.

Earth's gravitational pull – the bone thins and becomes fragile. This explains why bone loss associated with prolonged space flights is confined to the weight-bearing bones. Calcium leaches out of bone as it thins, which produces secondary complications because elevation of calcium in the urine enhances the risk of kidney stones. This demineralization leads to brittle bones (osteoporosis) and can increase the risk of fracture on return to Earth. Bone loss can be quite considerable on a long space flight. Astronauts lose around 1 per cent of their bone mass per month and ten months of microgravity can produce a reduction in bone mineral density similar to that which occurs between the ages of thirty and seventy-five on Earth.

Another serious consequence of prolonged microgravity is wasting of the weight-bearing muscles, due to reduced use. They shrink in size and strength, and become more susceptible to exercise-induced damage. Connective tissues also degenerate. These effects are observed primarily in the legs; the arm muscles seem less affected, probably because the arms are used almost exclusively for work in space. Atrophy of the leg muscles has serious implications, for it may impair the ability of the crew to evacuate the spacecraft unassisted following an emergency landing. The workload of the heart is also reduced in microgravity, both because of the reduced blood volume and because it no longer has to pump against the Earth's gravitational pull. Consequently, the amount of heart muscle declines and there is a measurable decrease in cardiac size after long flights.

To reduce bone and muscle loss, all astronauts must spend at least three or four hours each day doing exercises. This is not easy, for exercise in microgravity presents unusual challenges. To use a treadmill, astronauts must be tethered firmly to the surface to prevent them from floating off backwards when they try to run: usually they wear a flexible harness of elastic cords that anchors them down. Exercise bikes and rowing machines have also been used successfully in space. These may sport some unique modifications; the rowing machine, for example, needs no seat to support the astronaut's weight. Isometric exercises, which enable muscle loading without movement, are also used. One example is the rubber-band chest expander used on Earth. Weight-lifting is, of course, not an option. Cosmonauts also wear a 'penguin suit' for several hours a day; this is an elasticated garment that stresses the muscles and partially compensates for the lack of gravity.

Unfortunately, to date, it has not proved possible to exercise enough

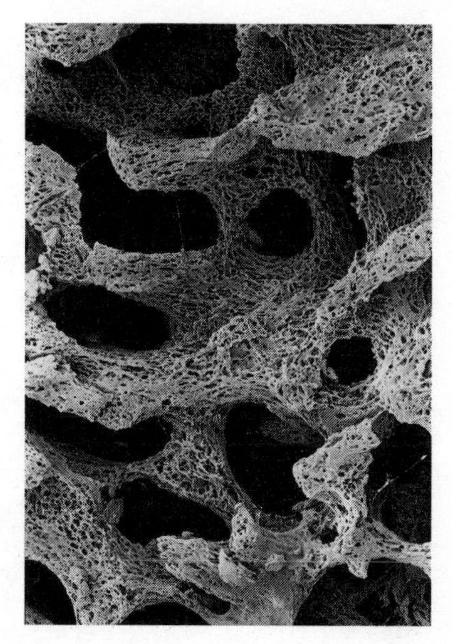

Left, a normal bone; right, a bone ravaged by osteoporosis. Bone is constantly being broken down and rebuilt. Normally, these processes occur at equal rates but in microgravity the cycle becomes unbalanced and the skeleton thins. A similar problem afflicts people in later life, being particularly prevalent in women after the menopause. Bone cells called osteoblasts make new bone, and cells called osteoclasts break down old bone. It appears that in microgravity the activity of the osteoblasts is depressed. This also occurs in many post-menopausal women due to lack of the female hormone oestrogen, and without replacement therapy they may lose up to three per cent of their bone mass annually.

to maintain the same degree of physical fitness as on Earth, nor to avoid loss of bone mass completely. Nevertheless, on any long-term space flight – such as a trip to Mars – astronauts must ensure that they keep to a regular exercise programme, because exercise is very effective at preventing muscle wasting.

Some of the long-term effects of space flight on the body can be simulated by lying down with the head lower than the body. Volunteers who have lain prone for a year in this way suffer bone loss and muscle wasting, and their heart works less efficiently. Likewise, bone loss occurs with ageing, probably because we no longer rush about so much and we take less regular exercise. Sitting at my computer writing this book does not provide the same stimulus to my bones as playing tennis (or even digging the garden).

Astronaut Michael Foale using the treadmill during the STS-45 mission. The webbing harness prevents him floating away.

Cosmic Radiation

Extraterrestrial radiation is a significant problem for astronauts. On Earth, the atmosphere and the Earth's magnetic field act as a shield so that, with the exception of visible light and radio waves, little radiation reaches the ground. In space, however, astronauts are continuously exposed to its harmful effects. There are three sources of extraterrestrial (cosmic) radiation: galactic rays, solar radiation and trapped belt radiation.

Galactic rays originate outside our solar system and rain down continuously on the Earth's atmosphere. They may arise from supernova explosions, or be emitted by other stars within the galaxy. They consist principally of protons (hydrogen nuclei) together with some alpha particles (helium nuclei), and they are highly energetic. When these primary particles hit the Earth's upper atmosphere, they collide with the nuclei of gas atoms to generate a shower of secondary particles which includes protons, neutrons, electrons, muons, pions and neutrinos. Primary galactic rays therefore do not penetrate the atmosphere and only a small fraction of the secondary particles that they generate ever reach the ground. In space, however, shielding must be provided if galactic radiation is to be prevented from reaching the astronaut.

The sun constantly emits a great stream of highly charged ionizing particles, consisting mainly of protons and electrons, that spirals radially outwards from its source at a speed of about 450 kilometres per second. Under quiet conditions, this solar wind typically contains around five particles per cubic centimetre by the time it reaches the Earth. On occasion, however, great flares erupt violently from the sun's surface, ejecting large amounts of material into interplanetary space. These flares have the force of a billion 1-megaton thermonuclear explosions and may spit out as much as 10 billion tons of particles in a few seconds. During such sunstorms, the amount of radiation reaching the Earth increases dramatically. Like the Earth's weather, it is very difficult to predict exactly when a sunstorm will occur. However, solar flare activity varies with an approximate eleven-year cycle and will be at its peak once again in 2001.

On Earth we live in a protected environment. The Earth's magnetic field shields us from cosmic radiation by trapping charged particles in a cloud around the planet. Huge numbers of these particles, predominantly high-energy protons and electrons, are concentrated in two distinct regions surrounding the Earth known as the inner and outer radiation belts, which were discovered by James Van Allen and his students in 1958. Each belt is shaped rather like a hollow doughnut (the technical term is a toroidal disc) and encircles the Earth with its central axis aligned with the equator. The closest that the inner belt approaches the surface of the Earth is about 300 kilometres and the outer belt may extend 45,000 kilometres into space, which is about one-sixth of the distance to the moon.

To understand why charged particles become trapped in the Van Allen belts, it is helpful to think of the Earth as a bar magnet with one end at the North Pole and the other at the South Pole. Lines of force flow from one end of the magnet to the other. Although these are invisible to us, they can be visualized using iron filings. Some bacteria and

Sunstorms and solar flares don't just affect astronauts and satellites: they can have dramatic effects on Earth. When charged particles emitted by solar flares reach the poles they excite gas atoms in the atmosphere, creating the spectacular display of light known as the *aurora borealis*, or Northern Lights. These flickering curtains of soft light are usually a greenish-yellow but on occasions can appear vivid purple, violet or blue. The colours are determined by the atoms with which the solar particles collide: excitation of oxygen produces green light, while red light is emitted by nitrogen. The *aurora borealis* is most spectacular at the poles because solar particles are swept here by the Earth's magnetic field, and then directed into the atmosphere along the field lines which enter or leave the Earth at the magnetic poles. If you are wondering whether there are any Southern Lights – there are, but they have received less attention as fewer people are around to see them.

animals that possess magnetic particles can also detect them. When a cosmic ray hits the Earth's magnetic field lines, the charged particles cannot cross them. Instead they are attracted to the poles, circling and gyrating as they go. At the poles, some of the particles escape and slip through into the Earth's atmosphere but the majority bounce back and return along the route they have already travelled, in reverse. This endless dance of the protons produces the Van Allen belts.

The Van Allen belts pose a serious problem to astronauts and satellites, for radiation doses can reach 200 milliSieverts per hour, and they limit orbits for space flight to below about 400 kilometres. In these low orbits the radiation doses are small, except for one place above the South Atlantic. The typical orbit used by the Space Shuttle passes through this South Atlantic Anomaly about six times a day and it is during this transit that Shuttle crews receive most of the radiation dose they collect in space. On the other nine orbits each day, the Shuttle does not pass through the South Atlantic Anomaly; consequently, extravehicular activity is confined to these orbits.

Although radiation levels in space are usually low, long-term exposure may damage the genetic material (the DNA) and thereby increase the risk of developing cancer, and if the DNA within the germ cells (sperm and egg cells) is affected, it may cause infertility or genetic abnormalities in the astronauts' children. The high-intensity radiation characteristic of a solar flare poses a more immediate danger, for it kills cells outright: death may occur within a few hours as a result of damage to the central nervous system, or within a few days because of the destruction of white blood cells or the rapidly-dividing cells that line the gut wall. If they were to be exposed to the full force of a solar flare, an astronaut would die of acute radiation sickness in hours. Moreover, because solar particle events can continue for many hours or even days, the effect of cumulative doses of lower-intensity radiation may also be considerable. Fortunately, solar flares are relatively rare events.

Although specialized equipment is usually needed to detect cosmic radiation, on some occasions it can be seen directly by the human eye. Buzz Aldrin and Neil Armstrong described seeing strange white star-like flashes during the journey to and from the moon in the lunar lander *Eagle.* Similar bright sparks and short streaks of light were also observed during translunar flights by astronauts on later Apollo missions, usually when their eyes were shut. These flashes occurred at a frequency of one or two per minute. They are believed to have been caused by galactic rays that penetrated the cabin walls and entered the astronauts' eyes, because comparable light flashes have been seen by human volunteers exposed to man-made particle beams. Some Shuttle crews have also reported seeing light flashes, which were particularly frequent in the South Atlantic Anomaly and lowest over the poles. These were presumably caused by trapped belt radiation. Precisely which part of the visual system is stimulated by ionizing radiation is still a mystery, but the consensus view is that the charged particles directly excite the retina.

Galactic rays and solar particles are highly energetic, so that it is difficult to provide adequate protection in a spacecraft. To be safe from the full intensity of a solar flare at least 10 to 15 gram/cm^2 of aluminium shielding is needed – more if the exposure is prolonged. Given the weight restrictions of space flight, it is not practical to provide this amount of shielding routinely, and therefore space crews are unavoidably exposed to cosmic radiation. All astronauts carry dosimeters to monitor the amount of radiation they receive. To date, this has been well within acceptable limits, although those astronauts who have spent a long time in space may have received a significant dose. Astronauts on the Apollo missions, for example, who spent less than two weeks in space, received only 6 Gray whereas the crew of Skylab 4 who spent eighty-four days in space were exposed to as much as 77 Gray. Some Russian cosmonauts who spent even longer in space received proportionally higher radiation doses. A number of them subsequently developed cancer, although it is not known whether this was as a consequence of radiation exposure. Because of the hazardous effects of cosmic radiation, it may be better to make long space journeys towards the end of one's life when there is less chance of developing cancer before dying of other causes. This is one rationale for sending older astronauts to Mars.

Supersonic aircraft, like Concorde, cruise so close to the upper edge of the Earth's atmosphere that they are not shielded from cosmic radiation and passengers and crew receive an average dose of around 10 microSieverts (μSv) per hour. During a trip from London to New York you would therefore accumulate 35 μSv. The maximal yearly radiation dose considered acceptable for the general public is 1 mSv (1000 μSv)[3] so it would take fourteen round-trips from London to New York to exceed this level. Put another way, the annual limit corresponds to 100 hours in the air. Clearly, both flight crew and frequent flyers might easily exceed this level. However, the recommended maximum for the occupational dose is higher: 20 mSv per year or the equivalent of more than five round trips to New York each week, which would not so easily be exceeded. Indeed, those Concorde crew who fly the maximum number of hours receive only 7 mSv/year.

Occasionally, however, a freak sunstorm can cause a rapid and substantial increase in radiation, which may reach levels as high as 25 mSv/hour. To cover such an eventuality, Concorde carries a radiation warning system, capable of detecting both neutrons and ionizing radiation (protons and so on), which is fitted in the passenger compartment and connected to a display unit in the cockpit. Should the radiation level exceed 0.5 mSv/hour, the crew are instructed to descend to lower altitudes where the aircraft will be protected by the atmosphere.

Despite the fact that Concorde has been flying for more than twenty years, this has never been necessary.

At an altitude of 10,400 metres, the cruising altitude of most passenger aircraft, cosmic radiation levels are about half those encountered by Concorde and supersonic military aircraft. However, the total exposure is about the same, because it takes much longer to reach your destination. This means that the radiation dose received is similar, whether or not you are fortunate enough to travel by Concorde. Interestingly, normal aircraft do not carry radiation monitors, partly for historical reasons and partly because the hazard is so low. Furthermore, nowadays, solar flare forecasting is good enough for there to be sufficient warning for aircraft to be grounded before the full force of a flare reaches Earth. (It takes around two days for solar particles to travel from the sun to the Earth.) The chief difficulty in forecasting sunstorms is that their course is hard to predict and many will miss the Earth – thus, space weather-forecasters face the inevitable problem of if and when to issue a warning.

It is incontrovertible that astronauts, flight crew and frequent flyers are routinely exposed to more cosmic radiation than the general

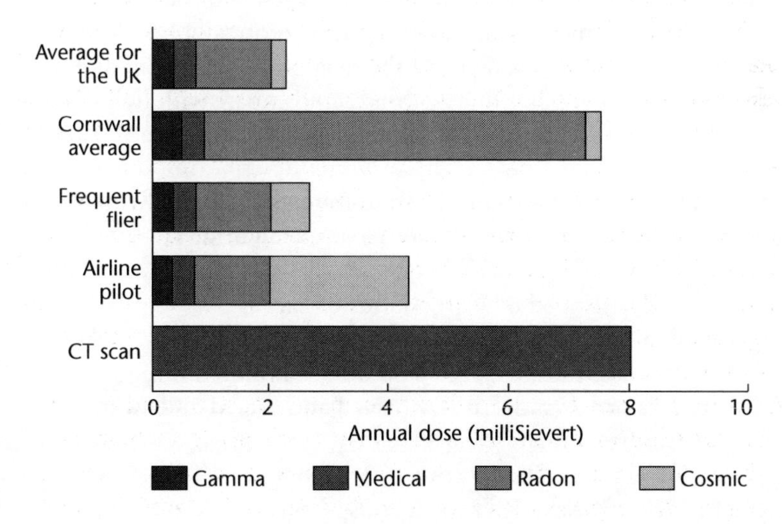

Sources of annual radiation doses for the UK population. The high radiation dose sustained by residents of Cornwall results because the granite rocks in this area give off radon gas, which can give rise to high radon levels indoors. A Londoner who spends a week in Cornwall actually receives more radiation than during a flight to New York. Significant radon concentrations are also detected in other areas of the UK.

population. Whether or not this leads to an increased risk of cancer is currently the subject of intensive investigation, but it is already clear that the risk must be very small and it must be set against the many advantages of air travel. It should also be considered in context. The one million inhabitants of La Paz, Bolivia (altitude 3900 metres), receive an annual dose of 2 mSv from cosmic radiation, around the same as the crew on long-haul intercontinental flights. People living in the southwest tip of Britain accumulate even higher doses, around 7 mSv each year, because of natural radiation from rocks. It is worth reflecting that while the flying hours of pregnant flight-crew are limited because of fears for their unborn child, pregnant women living in Cornwall are unavoidably exposed.

Venturing into the Void

The first man to venture into space itself, with only a suit to protect him, was Aleksei Arkhipovich Leonov of the Soviet Union. He spent twelve minutes outside his craft on 18 March 1965. The first spacewalk by an American, Edward White II, occurred a few months later. Today, astronauts from many nations have logged thousands of hours, both in space itself and on the moon. All of them agree that spacewalking is an exhilarating experience and that nothing can compare with flying in the void, in ultimate blackness, with the bright curve of the Earth turning slowly beneath you. None feels that words describe the feeling adequately; but Gene Cernan said that the following poem, written long before space flight was possible, came close to capturing its essence:

> Oh! I have slipped the surly bonds of Earth
> And danced the skies on laughter-silvered wings;
> Sunward I've climbed, . . . – and done a hundred things
> you have not dreamed of – wheeled and soared and swung
> High in the sunlight silence . . .
> And, while with silent, lifting mind I've trod
> The high untrespassed sanctity of space,
> Put out my hand and touched the face of God[4]

Spacewalking can be dangerous, for the slightest touch sends you spinning off into space. Consequently, astronauts are often tethered by

The flamboyant aviator Wiley Post (1899–1935) designed and successfully used the world's first pressurized flying suit. He used it to make a historic flight from California to Cleveland on 15 March 1935, taking advantage of the jet stream at high altitude to boost his speed to an average of 277 miles an hour. His pressure suit was constructed with three layers of material and was equipped with a helmet and an oxygen supply. Post found it very difficult to move when inflated, and eventually compromised by constructing the suit already in the sitting position, which allowed his hands and feet free movement. Wiley Post was no stranger to challenging flights – he made the first solo flight around the world in 1933, which took seven days and nineteen hours.

umbilical cords to the mother-ship and their spacesuits possess small rocket jets that allow them to manoeuvre in a vacuum.

Spacesuits evolved from the pressure suits used by pioneer aviators, such as the American Wiley Post, to help them attain high-altitude records. In the early days of flight, the cockpit was not pressurized and the only choice for a pilot who wished to venture to high altitudes was to wear a pressure suit. These early garments were subsequently developed by the military into full pressure suits for jet aircraft that fly above 12,000 metres. All early astronauts wore pressurized suits throughout the flight, as an emergency back-up in case the pressure inside the space capsule failed. In contrast, present-day astronauts wear normal clothing while in orbit, partial pressure suits during take-off and landing, and only don full pressure suits when venturing outside into the vacuum of space itself.

A spacesuit functions as a miniature personal spacecraft, providing physical protection, maintenance of pressure, an atmosphere, thermal stability, and – if it is to be used for an extended period – food, water and provision for removal of waste products. Spacesuits must also be flexible, tough and extremely resistant to solar radiation and micrometeoroids. To make it even more difficult for the designer, they must be lightweight because of the weight restriction imposed by the energy required to attain orbit. Early spacesuits, like those used by the Gemini astronauts, were supplied with oxygen by an umbilical line that tethered them to the main spacecraft. For the Apollo moonwalks, however, a self-contained life-support system was essential. The spacesuits worn outside the craft by NASA personnel today are highly complex and are known as extravehicular mobility units (EMUs). They have fourteen layers of material to protect the astronaut from the rigours of space, and a large back-pack containing the life-support system, which comprises tanks of cooling water, an air-conditioning system and gas tanks that hold sufficient oxygen for a nine- or ten-hour spacewalk. On Earth, the suit weighs a massive 113 kilograms; in space, of course, it has no weight at all.

The cabin pressures of the Space Shuttle and of *Mir* are maintained at that of Earth and the crew breathe an Earth atmosphere of 21 per cent oxygen and 78 per cent nitrogen. The EMU, however, is supplied with 100 per cent oxygen at a pressure of one-third of an atmosphere. Carrying pure oxygen (rather than an oxygen/nitrogen mixture) extends the time an astronaut can spend outside but means that the pressure

Astronaut Bruce McCandless II spacewalking without an umbilical tether, during the first occasion on which the nitrogen-propelled hand-controlled 'manned manoeuvring unit' was used. The Space Shuttle Challenger is reflected in his visor.

must be reduced to prevent oxygen toxicity (see Chapter Two). Exhaled carbon dioxide is filtered out by lithium hydroxide, activated charcoal removes other trace contaminants and water is removed by a dehumidifier. Oxygen is then added as needed and the gas recirculated to the suit.

Because the EMU is maintained at a pressure of a third of an atmosphere, an astronaut who simply donned his suit and immediately went outside would suffer decompression sickness. The symptoms of decompression sickness, commonly known as the 'bends', have been described in detail in Chapter Two in relation to diving. It results from the formation of bubbles of nitrogen gas in the bloodstream and body tissues. One way to eliminate the bends is to remove all nitrogen from the body and replace it with oxygen, because dissolved oxygen is used up by the tissues more quickly than it can form bubbles. Prior to extravehicular activity, therefore, Shuttle astronauts don face-masks and breathe pure oxygen. It is very difficult, however, for an astronaut to put on their EMU while continuing to pre-breathe oxygen. Since nitrogen rapidly resaturates the tissues, and a few breaths of nitrogen are sufficient to replace the oxygen that had taken so long to accumulate, the astronaut must hold their breath while the pre-breathing apparatus is replaced by the spacesuit life-support system and all nitrogen is flushed out of the system. This is not always easy. It is therefore common practice to reduce the cabin pressure and increase the oxygen level in the Shuttle atmosphere prior to pre-breathing, which considerably decreases the risk of nitrogen reabsorption. It also shortens the time required for 'pre-breathing': a pre-breathe of just thirty minutes is safe if the entire cabin is decompressed for twenty-four hours before the spacewalk, whereas without cabin depressurization pre-breathing must last for at least four hours.

Like a spacecraft, a spacesuit must operate under extremes of temperature for the temperature can climb to over 120 °C on the side lit by the sun, while on the dark side it may plummet below -100 °C (this must be the ultimate case of sitting in front of a roaring fire in an icy room). Furthermore, body heat and perspiration cannot escape through the skin of a spacesuit so potentially it can become very hot inside, especially if the astronaut exercises. Indeed, overheating was a considerable problem for the Gemini spacewalkers. Later spacesuits were provided with water-cooled underwear: the undergarments were threaded with a network of fine tubes through which water constantly circulated from a store carried on their backs. A similar system operates in the EMUs used by the Shuttle crew today.

A spacesuit must also allow the wearer to move around and work in space. This is not a simple matter to engineer. Astronauts working in space need to be able to bend their arms but a spacesuit must be strengthened with wire supports to resist the external vacuum[5] and a pressurized suit resists bending. Spacesuits therefore have integral joints at appropriate places, which operate rather like the exoskeleton of an insect. The lower part of the spacesuit, for example, is jointed at the waist, hip, knee and ankle just as the hard carapace of a beetle is jointed at key places. Even so, working in a spacesuit is difficult and extremely tiring, and rigorous training is essential. A further complication is that in microgravity the human body increases in height, the chest and head expand, and the thighs shrink, because the body fluids shift upwards from the legs to the chest. Designers of spacesuits must take these changes in shape into account. Several early astronauts had uncomfortable experiences because their suits became too tight.

Re-entry

The most dangerous part of space-flight is probably re-entry into the Earth's atmosphere and touchdown. It is no accident that President Kennedy's famous speech specified not only landing a man on the moon but also 'returning him safely to Earth'. Both physical and physiological problems beset the returning astronaut. Chief among these is the intense heat produced by the friction between the spacecraft and the Earth's atmosphere. The speed at which the spacecraft moves through the Earth's atmosphere strips electrons from atoms in the air and creates an ionized orange-red plasma around the spacecraft. The temperature can rise to a searing 1650 °C and a heat shield is needed to prevent the spacecraft from burning up and the astronauts inside from being fried. A further complication is that the upper layers of the atmosphere are not smooth but instead corrugated, like waves, so that severe vibrations may occur as the spacecraft bounces from ridge to ridge, skipping on the Earth's atmosphere.

Re-entry is especially dangerous for an astronaut who has spent a long time in space, because of the increased g-force associated with deceleration as the spacecraft enters the Earth's atmosphere. In early space-flights, these forces reached very high levels (+6g), but today's Shuttle pilots are exposed to a force that is only 1.2 times the Earth's gravity. Even this affects the astronaut's body. Because of the position of the Shuttle as it enters the Earth's atmosphere, the pilot experiences the g-force at an angle that makes it more difficult for the heart to pump blood back from the feet; and it may be sustained for as long as twenty minutes. This is a particular problem for astronauts who have spent a long time in space and whose bodies have adapted to microgravity. Their blood pressure may fall dramatically, leaving them dizzy and faint at the critical stage of landing. The British astronaut Michael Foale, who spent almost five months in the *Mir* space station, was strapped horizontally for re-entry on the Shuttle so that the g-force was applied from chest to back. Antigravity trousers, like those worn by military pilots, are also sometimes used to apply an external pressure and so assist the return of blood to the heart.

Coming Down to Earth

A common problem for astronauts on their return to Earth is that they are unable to stand without fainting. This condition, known as orthostatic intolerance, arises because weightlessness causes significant changes in the cardiovascular system. Freed from the stress of gravity, body fluids shift upwards, triggering compensatory mechanisms that reduce the fluid volume and promote its redistribution. These changes persist for some time after return to Earth. Although they have no noticeable effect while the astronaut is lying down, when they try to stand up, the blood supply to the head and brain is reduced, resulting in loss of consciousness. The crew of Soyuz 21, for example, had great difficulty in standing up without fainting for several hours after landing. Orthostatic tolerance is decreased after flights lasting as little as five hours. It returns to preflight levels within three to fourteen days after a short flight, but more time is required for recovery after longer space missions.

One reason astronauts suffer orthostatic intolerance on landing is that their blood volume is decreased, and the blood vessels in their legs do not constrict as much as normal, so that more blood collects in the legs under normal gravity. In addition, the nervous control of blood pressure seems to be impaired. People with low blood pressure, like myself, also find that when they stand up rapidly, they see black spots or a grey curtain in front of their eyes and feel dizzy for a few seconds.

The Soviets were the first to introduce measures to counteract changes in body fluid distribution. Cosmonauts used vacuum trousers at intervals during their flight, which applied external suction to drag the blood back to the lower body, and they drank about a litre of slightly salty water just before leaving orbit, in order to increase the volume of their body fluids.[6] Cosmonauts who practised these measures did not experience severe orthostatic intolerance on return to Earth. An exception was the crew of Soyuz 21, who skipped the orthodox programme because one of their colleagues developed a severe and persistent headache that necessitated a rapid return to Earth after forty-nine days in space. None of them was able to stand without fainting for several hours after landing. Tests on Shuttle astronauts have confirmed the beneficial effects of drinking saline solution before leaving orbit,

Earthrise over the moon – one of the most hauntingly beautiful of photographs – captured by Bill Anders in 1968 as his spacecraft encircled the moon. As Anders later remarked, 'we came all this way to explore the moon, and the most important thing is that we discovered the Earth'.

and both Russian and US crews now consume about a litre of water (or juice) and eight salt tablets just before their return. These measures are very effective against the orthostatic intolerance that results from short space flights but unfortunately do not seem to protect astronauts who have spent long periods in space.

Muscle wasting is another reason why astronauts who have spent a long time in space may find themselves unable to walk when they land on Earth. Space flight also makes muscle more susceptible to injury. Experiments on animals have revealed that muscles are not damaged by microgravity itself but rather by exercise on return to Earth. Usually, muscle recovers rapidly, so that walking is possible in a few days and muscle mass is restored in a few weeks. In contrast, bone loss may take many months to replace, depending on the duration of the space flight.

If you have ever spent any time at sea on a small boat, you will have found that you quickly accommodate to the swell of the waves, but that when you step on land once again the ground feels unsteady and it is difficult to walk. A similar instability afflicts astronauts on their return to Earth. About 10 per cent of the Shuttle crew suffer from Earth sickness on landing. They find it difficult to keep their balance or remain upright with their eyes closed and they complain of dizziness and nausea. Many more astronauts find that immediately after landing any movement of the head produces the illusion that the world, rather than the head, is moving. This suggests that signals from the receptors in the semicircular canals, which respond to linear acceleration, are reinterpreted during adaptation to microgravity and that they must be reprogrammed once more on return to Earth. For the first couple of nights after landing, many astronauts also find that when lying flat in bed they actually feel as if their head is about 30 degrees lower than their feet. Earth sickness passes within a few hours or days, but balance and coordination may take a week or two to return to normal. Interestingly, lunar gravity – one-sixth that of Earth – has less effect. Only three of the twelve men to walk on the moon reported any symptoms and these were so mild they could have simply been the result of excitement.

Whither Next?

When the lunar module *Falcon* of the Apollo 15 mission left the moon, its crew left behind a small plaque inscribed with the names of the fourteen astronauts and cosmonauts who had died in man's attempt to reach the moon, and a little figurine that is now known as the 'Fallen Astronaut'.[7] Without doubt, space is the most hostile of environments for humans. Yet none of these men died in space. They were killed when their capsule caught fire on the launch pad, when their rocket exploded during launch, or during re-entry and return to Earth. The most critical moments of space flight, like those of air travel, seem to be take-off and landing.

The inevitable question that surrounds manned space flight is, 'Is it worth it?' But when most people ask this question they are not referring to the cost in human lives. Space travel is extremely expensive. The Apollo missions to the moon cost the United States a massive 4.5 per cent of its annual budget. Felt to be necessary during the Cold War, support for the programme faded once the political imperative was removed, and the Apollo missions were terminated early. Shortly after leaving the moon, and docking with the mother-ship, Apollo 17 astronauts Gene Cernan and Jack Schmidt, the last men on the moon (so far), were horrified to learn of a statement by President Nixon that 'this may be the last time in this century that men will walk on the moon.'

His words were prophetic: no one has yet returned to the moon. The dream that became a reality for those of us over thirty today is once more only a dream. Today, our space programmes are far more modest. Robots, not humans, range over the surface of Mars. There is some justification for this strategy, for robots are far cheaper, require less support and do not endanger human lives. Yet I have no doubt that the same spirit that carried man to the moon, will eventually take the human race to Mars. I hope I shall be around to see it happen.

7

THE OUTER LIMITS

'The microbe is so very small,
You cannot make him out at all.'

HILAIRE BELLOC, 'The Microbe'

Black Smoker

WHEREVER MAN HAS ventured on the planet, other organisms have got there first. Even the most inhospitable parts of the polar regions, the deserts, the mountain tops and the depths of the oceans have been colonized. Few areas on Earth are so hostile that unicellular organisms cannot exist, and even in environments so extreme that humans cannot survive unaided, other animals live without difficulty. This chapter looks at the limits to life. It contrasts the narrow range of environments that people can endure with the much greater range that other organisms can tolerate and it considers how they survive deep inside rocks, in strong alkali, acid or salty lakes, in bogs, ocean depths or boiling mud pools.

In order to survive, animals like ourselves need water, oxygen and a supply of food. Bacteria can do without oxygen, and their food source may be very different from our own, but they still require water. They also need elements such as carbon, nitrogen, sulphur and phosphorus as building blocks for DNA and proteins. These elements are found in most places on Earth, but liquid water is less common. In the Atacama desert, reputed to be the driest place on Earth, it may not rain for years on end. Ice is no substitute for water, so the frozen wastes of the polar regions and the mountain tops are also deserts. Although some organisms can exist without water for extended periods, in a state of suspended animation, they are unable to grow or reproduce. Water is therefore the essence of life, the true *aqua vitae* sought so earnestly by the alchemists of old.

The Tree of Life

The tree of life has three main branches: the eukaryotes, bacteria and archaea. Eukaryotes, like ourselves, are made up of cells which possess a nucleus, that houses our DNA. All animals, all plants and many unicellular organisms are eukaryotes. Bacteria and archaea are single-celled organisms that lack a nucleus, but they are as different from each other as from eukaryotes and they each possess a unique set of genes. Surprisingly, the fact that the archaea constitute a distinct branch of life was only recognized recently, in the late 1970s, by the evolutionist Carl Woese. His insight was not accepted widely at first, and Woese felt bitterly disappointed that it was dismissed as nonsense, or simply ignored, in his home country, the United States. In the view of his

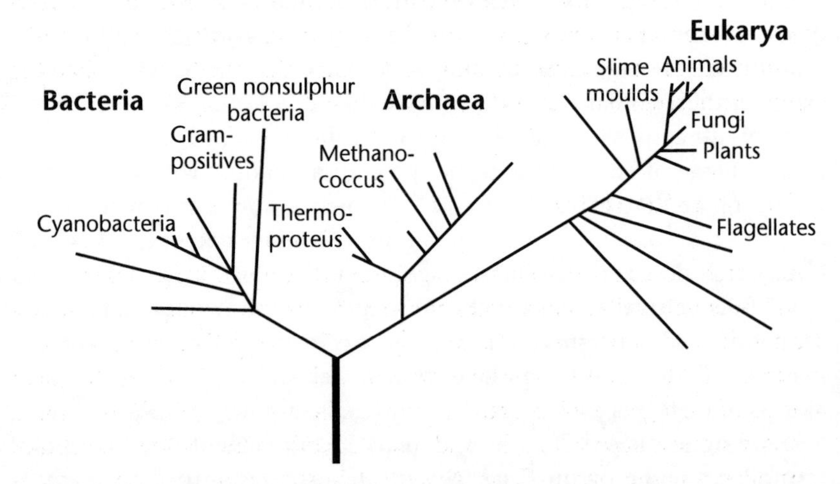

The tree of life suggested by Carl Woese is based on an analysis of the extent of relatedness between the genes of different organisms. It has three main branches: the eukaryotes, bacteria and archaea. The archaea and bacteria far surpass eukaryotes in both their number and diversity – there are thought to be as many as 10 million different species. Extrapolating backwards along the dendrogram suggests that the oldest life-forms were probably hyperthermophiles, that resembled the archaea that live today in the black smokers of the mid-ocean ridges, and the bubbling volcanic hot pools of Iceland and New Zealand. Scientists are still arguing about whether this is the case: no one really knows whether life began in a cauldron, in warm balmy seas, or even in icy waters.

opponents, the archaea were merely a specialized branch of the bacteria. Woese's retiring nature did not help to publicize his message but today his idea is well established. Conclusive proof of its validity came when the first complete genome sequence of an archaea (*Methanococcus*) was obtained in 1998, for its genes were found to be quite unlike those of bacteria, confirming that the archaea are indeed unique and showing that they have a closer relationship with eukaryotes than bacteria. The name archaea derives from the Greek *arkhaios*, and refers to their ancient origins, for it is thought that of all existing life-forms they most closely resemble the first cells.

The archaea and bacteria are the true masters of the extremes. They thrive in boiling water, in caustic soda lakes, in strong acid, in highly salty water, at enormous pressures and deep in the middle of rocks. Some, like *Deinococcus radiodurans*, can even tolerate extremely high levels of radioactivity.[1] Many can live without oxygen or sunlight, obtaining their energy from sulphur, hydrogen or by breaking down rocks, and they can digest almost anything including oil, plastic, metals and toxins. These microorganisms have enormous potential for use in environmental cleanup, pollution prevention, energy production and a multitude of other areas. Small wonder then, that they have excited the interest of scientists and industrialists alike.

Some Like It Hot

Although no multicellular animals or plants survive long above 50 °C, and no single-celled eukaryotes above 60 °C, some archaea and bacteria can tolerate temperatures close to boiling. Thermophiles live at temperatures of 50 °C and hyperthermophiles above 80 °C. They frequent areas of high geothermal activity, such as the hot springs of Iceland and Yellowstone National Park, and the volcanic vents known as black smokers on the ocean floor. The most heat resistant found so far is *Pyrolobus fumarii*, which lives at temperatures of 113 °C in the walls of the black smoker chimneys and stops growing when the temperature falls below 90 °C because it finds it too cold. No one knows what the upper limit to life actually is, but most scientists suspect it lies around 120 °C.

Black smokers were first discovered in 1977 by scientists from the Woods Hole Oceanographic Institute, off the coast of Ecuador.

Gliding along the ocean floor in the submersible *Alvin*, at a depth of around 2500 metres, they crested a ridge and were greeted by the most extraordinary sight. Clouds of black smoke billowed forth from a forest of underwater smokestacks, as if Vulcan and Neptune had collaborated to create some strange industrial complex of submarine satanic mills. And, unlike the sparsely populated ocean depths to which the scientists had become accustomed, this was an oasis of life that teemed with many different kinds of animals.

Exploring black smokers can be hazardous. When scientists first inserted a temperature probe into the water ejected from the chimneys they found that their instrument suddenly ceased to function – the probe had been burnt off within a few seconds. Once the cause of the problem was recognized, there were immediate concerns about the submersible itself for, at that depth, its Plexiglass windows lose their strength at a mere 90 °C. Nor was this concern unjustified. Occasionally, submersibles return from expeditions to black smokers with their outer fibreglass shell charred black.

Black smokers resemble underwater geysers, spewing out superheated water laced with minerals from volcanic vents in the ocean floor. At the mid-ocean ridges, hot magma from the Earth's core comes to the surface, pushing apart the tectonic plates and hardening when it cools to form new ocean floor. Cold sea water percolates down through cracks in the sea-bed and is heated as it sinks by the hot magma. As it descends, it gets progressively hotter but it does not boil because of the immense pressure.[2] Eventually, the superheated water gushes back up to the surface laden with dissolved minerals and metal sulphides and belches forth from vents on the ocean floor at over 350 °C. As soon as it hits the cold sea water, the dissolved metals and minerals precipitate out, forming a smoky black plume that rises 100–300 metres above the ocean floor and subsequently solidifies into rocky chimneys, as much as 5 metres high. One of the tallest, an impressive 6.8 metres, is affectionately known as Goliath.

The water around a black smoker teems with archaea that thrive in the superheated waters and feed on the cocktail of sulphur compounds and minerals (manganese, iron and sulphide) ejected by the vent. Clouds of these chemosynthetic organisms, looking like snow and known as 'flock', surround the vents. They support a unique ecosystem. Colonies of tube worms wave their tentacles like grass in the warm currents. Some have small, delicate, etiolated shells encircled with ridges, while others are up to 4 metres long. The Pompeii worm, one of the most heat tolerant, lives in tubes attached to the side of the black smokers themselves. While its head emerges into water of a comparatively pleasant 20 °C, its tail is subjected to a scorching 80 °C. Billions of shrimp swarm around the vents, dancing a thermal tightrope: too close to the smoker and they are boiled alive, too far away and they freeze or starve to death. Sea anemones, spindly-legged crabs and giant clams (almost a foot long) decorate the sea bed. Microbial mats coat the surface of mussels. Fish forage among the chimneys. A strange and beautiful orange creature trailing long filaments, that looks like a dandelion gone to seed, sails past – not a flower, but a colony of animals related to jellyfish, like the Portuguese Man-o'-war.

No sunlight penetrates these depths, so all life is ultimately dependent on the chemosynthetic abilities of the archaea and bacteria. These use hydrogen sulphide as a fuel supply, oxidizing it to water and sulphur. While some creatures feed directly on the clouds of archaea, others survive by forming an even closer relationship. One of the most extraordinary is the tube worm *Riftia pachyptilia* which has a soft white cylindrical body, thick as a child's arm, that is tipped with crimson gills. It has no gut or excretory system for it does not feed in the conventional sense, but instead obtains its energy from a symbiotic partnership with chemosynthetic bacteria. The interior of the tube worm's body is filled with a structure known as a trophosome, or feeding sac. Thousands of sulphur bacteria live within each of the cells of the trophosome. The brilliant blood-red gill plumes of the tube worm extract oxygen and hydrogen sulphide from the surrounding water. These combine with a specialized form of haemoglobin in the worm's circulatory system and are transported to the chemosynthetic symbionts that inhabit the worm's central cavity. The bacteria use the oxygen to split the hydrogen sulphide into water and sulphur, releasing energy in the process. The sulphur is left behind, forming a solid yellow deposit that accumulates in the tube worm's body throughout its life. The energy is used to convert inorganic compounds into nutrients, such as amino acids and carbohydrates, that the bacteria then share with their host.

The first thermophilic microorganisms were not discovered in the black smokers but in the superheated geothermal waters of Yellowstone National Park in Wyoming. Yellowstone is a surreally beautiful world of fire and water. Hundreds of hot springs and boiling pools litter the

Hot sulphur pool in Yellowstone National Park

landscape, fringed with gelatinous pink and purple mats of microorganisms. Giant columns of water are shot into the air with such force that the ground quakes. Steam hisses and roars from fissures in the ground like angry dragons. Bubbling mud pools and explosive mud pots grumble away comparatively quietly. Water cascades over rainbow-coloured rocks, stained by the bacteria and archaea that coat their surfaces. The air is thick with the stench of rotten eggs – this is hydrogen sulphide, a malodorous toxic gas that burns the throat and makes it hard to breathe. The superheated pools are scalding hot, but they are far from barren. Dip a stick into the water and it becomes coated with gooey black slime, a glutinous mass of archaea and bacteria that love the heat.

Thomas Brock and his wife Louise were the first to explore whether there was life in these boiling waters. One summer in 1965, they spent a working vacation in Yellowstone and isolated the first hyperthermophilic organisms in a run-off channel from a hot, acid, sulphur-rich pool. This was *Sulpholobus acidocaldarius*, which prefers a temperature between 60 and 95 °C. Another of their finds was *Thermus aquaticus*, destined to become the star of the biotechnology industry. The Brocks' discoveries initiated the study of extremophiles, generated a new breed of prospectors – the microbe hunters – and launched what was to become a million-dollar industry. They also provided a perfect excuse for microbiologists to visit the more remote and bizarre regions of the planet in search of microorganisms previously unknown to science.

At the time Thomas Brock isolated *Sulpholobus*, the scientific dogma held that nothing could live above about 50 °C, which is perhaps why no one had thought to look for life in such an extreme environment before. The key to Brock's success was that he cultured the bacteria he collected at a temperature as high as that at which they lived normally. A less astute scientist might have been tempted to try a lower temperature, in the mistaken belief that the bacteria would grow better. They would have found nothing, for *Sulpholobus* is an obligate thermophile. The isolation of the first extremophile, like so much innovative science, required a combination of acute observation and a willingness to consider the heretical. The memorable claim of the White Queen in *Through the Looking-Glass* that she 'sometimes believed as many as six impossible things before breakfast' is not bad advice for a scientist.

Multicellular animals cannot match the heat tolerance of thermophilic archaea and bacteria. One of the multicellular record holders is the Pompeii worm, which lives in a tube cemented to the walls of a black smoker and tolerates a thermal gradient of 20–80 °C from head to tail. Another is the silver ant of the Sahara; it is able to forage in air temperatures close to 55 °C, but only for brief periods before it must retreat underground to cool off.

Heat tolerance has evolved to enable an organism to exploit an ecological niche inaccessible to others. But it can also be used as a weapon. The Japanese honey bee (*Apis cerana japonica*) employs its body heat as a defence against the predatory hornet wasp (*Vespa mandarinia japonica*), which is more sensitive to thermal stress. When the colony is attacked, hundreds of honey bees mob the invader. The temperature at the centre of the swarming mass of bees quickly rises as high as 48 °C, which is lethal to the wasp, but not to the bees. Quite simply, the invader is cooked to death.

Most cells die at temperatures over 50 °C because their proteins do not like being overheated. The molecular vibrations that occur when proteins are exposed to heat shake them apart, causing mature proteins to unravel and preventing new ones from folding correctly when they are being made. This denaturation is dangerous, for the protein can no longer do its proper job. Structural proteins are degraded, and enzymes fail to catalyse biochemical reactions. The importance of correct protein folding has become abundantly clear to the British public in recent years, for bovine spongiform encephalitis, more commonly known as BSE or 'mad cow' disease, is the result of a rogue form of protein that folds aberrantly and is able to seduce others to adopt its ways. For reasons still not understood, the incorrectly folded form of the protein is toxic and causes neuronal death.

Thermal damage to proteins is not easily reversed. Egg white that has been cooked remains hard, white and rubbery and does not revert to its original translucent, glutinous form on cooling. A cold cooked steak may be unappetizing but it is still clearly a piece of cooked meat, its muscle structure irrevocably destroyed by cooking. Cells are able to repair less severe damage, however, using heat-shock proteins. These molecular chaperones restore order by helping proteins refold correctly. Proteins that are irreversibly damaged and cannot be refolded are tagged and consigned to degradative pathways that break them down and recycle their constituent amino acids for further use. Thus, heat-shock proteins act as a kind of biochemical fire brigade.

Proteins are formed from a linear string of amino acids but – like a bead necklace dropped on the floor – they fold up into far more complex shapes. Sometimes two or more protein chains may link up to build a much larger molecule: insulin, for example, is composed of two protein subunits and haemoglobin consists of four. The three-dimensional shape of a protein is critical. Signalling molecules must dock snugly with their target receptors, enzymes must embrace their substrates in precisely the right way, structural proteins must lock themselves tightly in position. The sequence of a protein dictates the way it folds up, but inside the cell this process is hindered by the very high concentration of other proteins. Such molecular overcrowding means that a protein may accidentally form bonds with adjacent unrelated proteins instead of with itself. Chaperone proteins have evolved to ensure that each protein makes the right connections, rather like their human Victorian counterparts. They provide proteins with a helping hand even at normal temperatures, but at high temperatures their numbers increase dramatically. In fact, it was the observation that the manufacture of chaperone proteins was stimulated by heat that led to them being christened 'heat-shock' proteins. We are left with a final puzzle, however: what ensures that the heat-shock proteins fold up correctly when things get hot?

The heat tolerance of hyperthermophiles is not just due to the activity of chaperone proteins, however. Many other enzymes and structural proteins – even the protein-synthesizing machinery itself – exhibit unusually high heat resistance. Despite being far more heat stable, it turns out that several hyperthermophile enzymes are very similar, at the amino acid level, to our own. Just a few critical amino acids seem to account for their extraordinary heat resilience.

Acid Heads

One dark night, trying to replace the battery in my car, clutching a torch in one hand and a spanner in the other, I inadvertently dropped the spanner. It lodged itself across the terminals, shorting the battery and causing it to explode spectacularly, showering me with acid. Fire stung my face and hands as the acid etched my skin. In the ensuing panic to rinse my eyes, the acid splashes on my trousers went unnoticed – until the next day when, wandering round the town, my jeans crumbled into holes and fell apart.

Like the cotton fibres of my trousers, the organic compounds in our cells are destroyed by acid. Acid baths strip flesh from bone and are used to bleach skeletons for anatomical display. They feature in detective novels as a lurid, if unconventional, means of disposal of an unwanted body. And they are not confined to fiction. The infamous 'acid bath murderer', John Haigh, who killed at least six people in Britain during the 1940s, used a bath of sulphuric acid to dispose of the bodies. One incriminating piece of evidence gave him away; a set of false teeth made from acrylic resin that did not dissolve. Acids are also used for more beneficial purposes. As the advertisements remind us, bleach, a mild form of hydrochloric acid, kills many pathogens. Acid is simply not good for most organisms.

The acidity or alkalinity of a solution (its pH) is related to the amount of hydrogen ions it contains. The more hydrogen ions present, the more acid it is and, conversely, the fewer hydrogen ions, the more alkaline (basic) the solution. The pH is defined as the negative logarithm of the hydrogen ion concentration. This means that an acid solution, which has a high concentration of hydrogen ions, corresponds to a low pH. Conversely, an alkali solution has few hydrogen ions but a high pH. This inverse relationship can be somewhat confusing at first, but in recent years pH has become a household term. Soaps and shampoos – and even some soft drinks – proclaim themselves 'pH balanced'. Gardeners, too, must be aware of the pH of their soil, for acid-loving plants like heathers and azaleas cannot tolerate alkaline chalky soils, while Cheddar pinks love limey soils and die when grown on acid ones. It is also worth remembering that pH is a logarithmic function, which means that a change of one pH unit corresponds to a ten-fold difference in hydrogen ion concentration. Thus vinegar (with a pH of 2) contains almost a billion times more hydrogen ions than ammonia (pH 11).

Most cells prefer an environment that is close to neutral pH (7.0), where the concentration of hydrogen ions is exactly balanced by an

Helicobacter Pylori, the Bacterium that Causes Stomach Ulcers

As recently as 1980, stomach ulcers, the hallmark of the ambitious executive, were generally believed to result from excessive acid secretion precipitated by stress. Two Australian pathologists, Robin Warren and Barry Marshall, were not so sure. They had found spiral-shaped bacteria in stomach washes from patients with ulcers or gastritis (a chronic inflammation of the stomach). The key question was whether these bacteria were contaminants, or actually lived in the stomach. Having determined their indigenous origin, they then had to confirm that *H. pylori* was indeed the cause of gastritis and ulcers rather than merely a harmless bacterium coincidentally found in association with the disease. Two intrepid volunteers, one of them Marshall himself, swallowed a concoction containing the bug to find out. Sure enough, they became ill with gastritis.

Warren and Marshall's experiments produced a radical change in medical thinking almost overnight. Ulcers, it became apparent, are not simply the result of extreme acid production by the stomach, but are instead due to a bacterial infection. It is postulated that the presence of *H. pylori* in the stomach wall causes an inflammation that eventually leads to tissue destruction and ulceration. Medical practice was also revolutionized. It became clear that drugs which inhibit acid formation provide only temporary relief because the bacteria remain. A course of antibiotics, however, eradicates *Helicobacter pylori* for ever. It is the difference between a cure and a equal concentration of hydroxyl ones (a hydrogen ion and a hydroxyl ion combine to make a water molecule). Cells are also very sensitive to small changes in pH, which is why the pH of human blood is tightly regulated. Its normal value is around 7.4, and an increase above 7.7 or a decrease below 7.0 is incompatible with life.

Astonishingly, some archaea and bacteria prefer it very acid, or even very alkali. Acidophiles like to live below pH 5. They inhabit the hot springs of geothermal areas, where sulphurous gases dissolve in water to produce sulphuric acid, and live in the acid waters that leach from the slag-heaps that litter the sites of old coal mines. Others live in vinegar and lemon juice, which is why these solutions deteriorate over time. One of the most fascinating is *Thiobacillus ferrooxidans*. It uses

therapy – between eliminating the disease and simply treating its symptoms.

Marshall and Warren's findings have enormous public health implications, for it is estimated that around one-third of the world's population is chronically infected with *H. pylori*, although not all of us will develop the disease. They are also of consequence to the pharmaceutical industry. Zantec, which suppresses acid secretion in the stomach, made a fortune for Glaxo and is still one of the world's best-selling drugs. One might imagine that novel antibiotic treatments would reduce the market for such inhibitors of acid secretion substantially. Fortunately for the drug companies, this is not the case. Antibiotics turn out to be more effective in combination with a suppressant of acid secretion (although expensive antacid drugs are not actually needed – bismuth does the trick just as effectively).

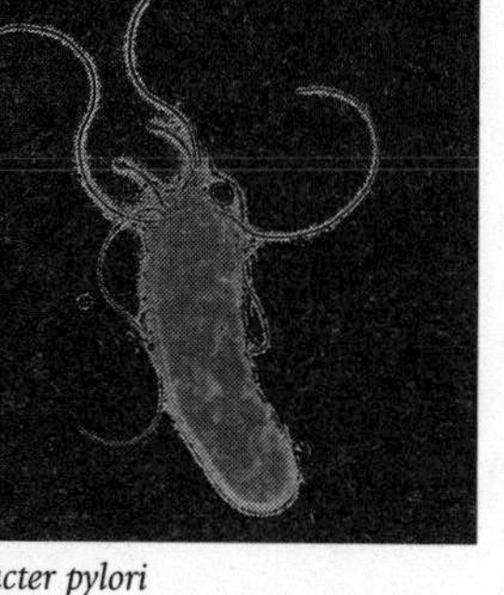

Helicobacter pylori

Despite living in the stomach, where the pH is as low as 2, *H. pylori* is not an acidophile. It actually prefers a neutral environment and although it can tolerate acid for a short time, it eventually dies if exposure is prolonged. Behavioural rather than physiological adaptations enable its survival in the stomach. It lurks within the mucous layer that coats the stomach wall and protects the cells lining the stomach from being etched away, and, as an additional preventative, it envelops itself in a cloud of higher pH by secreting the enzyme urease.

carbon dioxide, oxygen, sulphur and ferrous iron to make energy and in the process produces sulphuric acid and ferric iron salts, which stain the streams that drain old coal mines a bright-yellowish brown and turn the water very acid (as low as pH 2). Both the acid and the dissolved metals are toxic to most aquatic life-forms. But *T. ferrooxidans* is even more formidable, as its alternative name, *T. concretivorans*, hints. It has a particular predilection for low-grade concrete which is high in sulphur, particularly if the concrete is reinforced with iron rods. To the engineer's consternation, the bacterium can produce so much sulphuric acid that it rots concrete, leading bridges and flyovers to collapse, and tower blocks to crumble. It took some time before it was recognized that concrete rot is due to a bacterial infection, for the microbial density is very low – one bacterium needs fifty times its own weight in iron for a single cell division.

Acidophiles not only tolerate low pH, they actually prefer it. *Sulpholobus*, for example, grows best at pH 2. This is fortunate because it actually produces sulphuric acid as a waste product of its metabolism. The optimum pH for other bacteria is even lower. The current record is held by microbes of the *Pircophilus* species which are happiest at pH 0.5, stop growing when the pH rises above 3 and disintegrate at pH 5. Some fungi and algae also tolerate acidic environments, and grow in weak sulphuric acid.

Acid destroys DNA and proteins. This raises the question of how acid-loving archaea and bacteria manage to tolerate a pH as low as 0.5. The answer to this conundrum is not really understood, but it is thought that they probably survive by keeping the acid out and that as fast as hydrogen ions enter their cells they pump them out again, or transform them into water by combining them with hydroxyl ions. But the proteins in the cell membrane, such as those that act as acid pumps, must be able to tolerate a pH of 0.5, because their outer surfaces are exposed to the extracellular acid environment. So the question is merely shifted back one step – why does acid not denature these proteins? No one yet knows, but many people are currently trying to find out.

Basic Needs

A string of alkaline soda lakes winds its way through the great Rift Valley of East Africa. Beautiful but inhospitable, these lakes are saturated with caustic soda. Sodium carbonate leaches from the surrounding volcanic rocks, turning the water that feeds the lake alkaline by scavenging hydrogen ions to produce sodium hydroxide (caustic soda). In the hot sun of the tropics, evaporation from the lake surface can be intense and exacerbates the alkalinity. The water in some of the Rift Valley lakes is undrinkable. Others are so saturated with soda that it forms a sparkling white encrustation at the lake rim, and the air is so caustic that it burns the throat and stings the eyes. Even fiercer conditions are found elsewhere. The soda lakes of southern Africa and of the Andean plateau may dry out completely, leaving spectacular glittering white deposits. In geological formations at Maguian in Jordan, the groundwater is so caustic (pH 13) that it will dissolve rubber boots. Yet, even here, life flourishes.

Many species of algae, bacteria and archaea thrive in the Rift Valley soda lakes, and they support a thriving population of brine shrimps. Millions of flamingos flock to the lake shores to feast on these small shrimps and on the cyanobacteria, red algae, and invertebrates that live in the surface waters or in the mud at the bottom of the lake. Rafts of these beautiful birds congregate along the lake shores, so that from the air the blue waters seem to be fringed with pink. Carotenoid pigments contained in the red algae and brine shrimps on which they feed stain the flamingos' feathers their characteristic colour. The flamingo is one of the few birds that can tolerate the caustic conditions of the soda lakes, but it too may run into problems.

The vast soda pans of Lake Natron in Kenya are so caustic that few animals dare venture there. Safe from predation, flamingos nest in large numbers during the cooler season, when large shallow alkaline lagoons lie scattered over the pans. But the lakes do not last long. As the dry season advances and the heat intensifies, the water evaporates and the alkali becomes more concentrated. At some point the water can no longer hold all the dissolved sodium hydroxide, which precipitates out of solution. It encrusts the legs of the flamingos, forming heavy anklets that weigh them down. The birds must abandon the lake before this happens. Leave it too late and they are chained to the lake, doomed to die a painful death from dehydration. This rarely happens in the case of adult birds, for they can fly to safety. But the chicks and juveniles, whose flight feathers are not fully developed, must walk out across the shrivelling lethal lake. For them, timing is all.

Like acid, alkali rots flesh and fibre. Accidentally spill some caustic soda (sodium hydroxide) on your clothes or skin and you will be painfully aware of its effects. Lime (calcium oxide) is a caustic white rock, produced by heating limestone, that combines with water to produce the highly corrosive calcium hydroxide. In the Middle Ages, lime pits were used to remove hair or fur from skins, and to bury the victims of plague. They are still used today when earthquakes or other natural disasters kill so many people that their rotting bodies become a health hazard.

Alkaliphiles take caustic conditions with impunity, favouring habitats with a pH above 9. This poses a problem because ribonucleic acid – the molecular messenger that carries the genetic information from the DNA in the nucleus to the protein-manufacturing factory in the cytoplasm – breaks down around pH 9. Consequently, no alkaliphile can afford to let its internal pH rise too high. They manage to keep their intracellular pH low by actively taking up hydrogen ions from the environment to raise their concentration in the cell to some near-normal level (remember that the pH and hydrogen ion concentration are inversely related).

A Salty Tale

Most organisms cannot tolerate salt, which is why salt was used as a preservative long before the advent of icehouses or refrigerators. Halophiles, however, flourish in highly salty seas such as the Dead Sea and the Great Salt Lake in Utah. Salt lakes arise when more water is lost by evaporation than enters from feeder streams. In hot environments, therefore, they may arise transiently during the summer months. Because salt water is heavier and sinks, salt lakes tend to be more saline at the bottom and fresher on top. Some salt lakes are also highly alkaline, and their inhabitants must be adapted to both high salinity and high alkalinity.

The briniest sea of all is the Dead Sea, with a salt content of 28 per cent, ten times that of the ocean. This is about as much salt as water can hold without it precipitating out. The density of the Dead Sea is so high, in fact, that it is possible to sit in the water and read the newspaper, as many postcards and holiday snaps testify. Located in the deepest land valley on the Earth, the Dead Sea lies 400 metres below sea-level in the Jordanian desert. The intense desert sun causes substantial evaporative water loss, so that despite being fed by freshwater streams its salinity does not decrease. The Dead Sea is a most inappropriate name, for it is very far from dead. Large populations of algae, bacteria and archaea thrive in the salty waters. Most are obligate halophiles and they cannot tolerate salt concentrations of less than 15 per cent. Some have spectacular colours, like the red halobacteria which on occasion proliferate so much that the sea turns the colour of blood.

Left to itself, Nature has a tendency to equalize things out.[3] Tip a glass of sea water into one of fresh water and eventually you will get a homogeneous solution. Because cell membranes are not completely impermeable to water, a cell that is placed in a highly salty solution will shrink, as water leaves the cell to equalize the concentration in the internal and external solutions. Consequently, the cell becomes dehydrated. This is the problem that confronts halophiles. Many cope with it by increasing the salt concentration inside their cells to that of the surrounding environment. Some of them, like *Halobacterium salinarium,* concentrate potassium chloride to very high levels – more than 200 times that in our own cells. Others adopt a different tactic and produce organic solutes that help retain water within their cells. Of course, this strategy only substitutes one problem for another, for their enzymes must then be able to cope with the high intracellular salt levels. How they do so is still an open question.

Archaea and bacteria are not the only inhabitants of salt lakes. Some algae can also survive there. They colour the water in brilliant shades of red, blue and green, and serve as food for a small crustacean, the brine shrimp *Artemia salina,* that also tolerates extremely salty conditions. *Artemia* is one of the few multicelled creatures that live in the Great Salt Lake of Utah. At certain times of the year, its eggs litter the surface of the lake, forming a dust of minute brown particles that is blown about by the wind. The eggs are extraordinarily tough, being resistant to drought and salt, and they can remain in suspended animation for a considerable period until reawakened by immersion in water.

Life in the Rocks

There is an abundance of life underground in literature and legend. Dwarves mine precious metals, fairies and hobbits live 'under the hill', fabulous dragons stand guard over caves heaped with treasure. Here too lie the mines of Mordor and the homes of trolls. Many ancient peoples believed that the souls of the dead also lived in the bowels of the Earth – logically enough, since there was clearly not enough room for them all above ground. When Orpheus went in search of his dead love Eurydice, he had to enter the vast cavern beneath the earth where the god Hades reigned. The kingdom of the Mesopotamian god Nergal, and his retinue of demons and devils that eternally destroyed one another, was also located underground.

For many years, biologists thought that life deep beneath the earth was no more than myth, and that living organisms petered out a few metres below the surface. But not any more. Unbelievably, microorganisms can survive deep inside rocks far beneath the Earth's surface, where there is no light or oxygen and the pressure can be considerable. They were first discovered in the 1920s in ground-water samples collected from oilfields hundreds of metres below ground. Although these reports were initially dismissed as artefacts, due to contamination introduced during the drilling process, the ability of microorganisms to live in deep rocks is now established fact. In 1992, Texaco was exploring for gas and oil in sedimentary rocks 2.8 kilometres below Taylorsville Basin in Virginia, which provided scientists with an opportunity to look for subsurface life. Even when using stringent sterile sampling conditions to prevent contamination by surface organisms, bacteria were found. Moreover, they were mostly previously unidentified species that required no oxygen but instead used manganese, iron and sulphur to oxidize ancient organic matter in nearby rocks for energy. They were also heat tolerant, for the temperature of the rocks in which they lived was over 60 °C. One species was christened *Bacillus infernus* because of its habitat.

Microorganisms have now been found far beneath the surface of the Earth and the ocean floors, in both sedimentary and igneous rocks. Sedimentary rocks were laid down on the surface of the Earth and then sank, so that it is possible the types of microorganisms found in these rocks are as old as the rocks themselves – trapped there since the rock was formed millions of years ago. This type of rock is riddled with fine pores, so microorganisms live everywhere. But their density is low. When cultured in the laboratory less than ten bacteria are found in each gram of rock,[4] far fewer than the 1 billion bacteria that live in every gram of garden soil. Igneous rocks such as granite and basalt are formed by solidification of molten volcanic magma. Because these

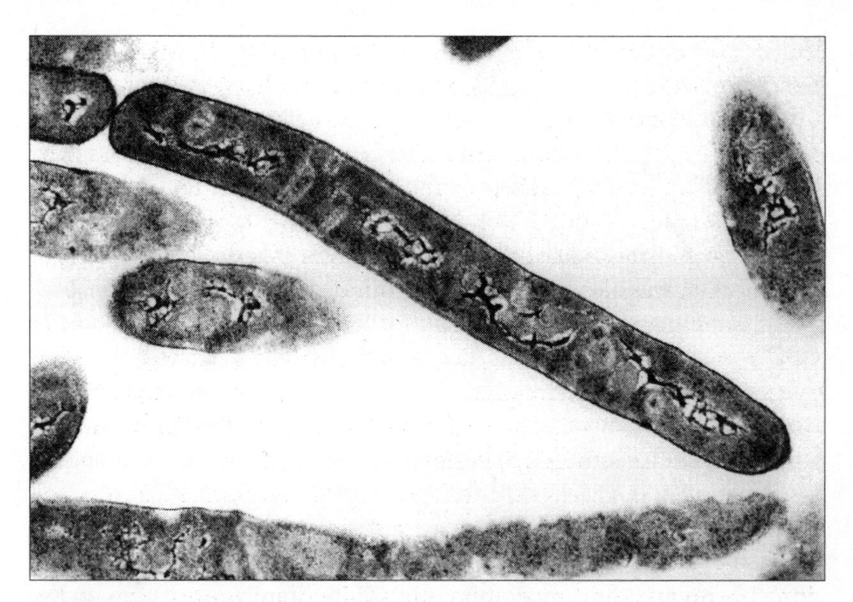

Bacillus infernus, the bacterium from Hell, lives 2.7 kilometres below the surface of the Earth where there is no oxygen, no organic food, the pressure is several hundred atmospheres and the temperature is over 60°C.

rocks are fused solid, most bacteria are found within fine cracks but, in some cases, the microbes create their own tunnels by dissolving the rock itself. All microorganisms within igneous rocks must have arrived once the rock had cooled, carried there by water seeping slowly down from the surface over many thousands of years.

In the laboratory, the bacteria isolated from samples taken from the Taylorsville Basin grew extremely slowly. By calculating the effect of bacterial respiration on the rocks within which they live, Tullis Onstott of Princeton University and his colleagues estimated that the average doubling time of the bacterial population in their natural environment was extremely slow (several thousand years). Deep in the rocks, it seems the microbes are merely surviving rather than multiplying. Because the rate of evolution is largely dictated by the rate of reproduction, these bacterial species may have remained largely unchanged for millions of years, entombed since the rocks were laid down at least 80 million years ago. Jules Verne's vision of primordial species still clinging to existence in the bowels of the Earth is not so far-fetched. He may have been wrong on the question of scale, but his idea of living fossils deep beneath the ground appears to have been remarkably prescient.

One problem for life below ground is that organic matter is extremely scarce. In rock cores taken from basalt rocks of the Columbia river there is too little organic matter to support life. Yet a surprising abundance of microbes is found. It turns out that the bacteria are feeding on the rock itself. As the rock weathers, hydrogen is released; this is used by bacteria to convert dissolved carbon dioxide into biomass, producing methane as a waste product. Weathering of rocks is usually attributed to chemical processes that slowly degrade the surface layers. Some scientists, however, have suggested that microorganisms themselves may play an important role in weathering, nibbling away at the surface of the rock over the aeons, extracting minerals and depositing elements in the Earth's crust.

The gold mines of South Africa are the deepest on Earth, lying 3.5 kilometres below the surface, where the pressure within the rock is 400 atmospheres and the temperature is 60 °C. Yet here, too, archaea thrive, as Tullis Onstott and Tom Kieft (from the New Mexico Institute of Mining and Technology) found when they visited the mines in 1997. The ultimate depth at which life can exist depends not on the weight of the rocks above, for unicellular organisms can withstand high pressures with relative impunity, but on the temperature of the rocks within which they live. The temperature rises towards the interior of the Earth – about 11 °C for every kilometre – because of the heat generated by the radioactive decay reactions in the planet's core. Deep-dwelling organisms must thus be hyperthermophiles. Assuming an upper limit of 120 °C for life, archaea will be confined to the upper 5 kilometres or so of the Earth's crust.

Troglodytes

Even bacteria locked inside rocks are perhaps not so strange as the unique sulphur-based ecosystems found in caves. The Movile caves in Romania were formed more than 5.5 million years ago and rockfalls closed their entrance. Sealed away from the outside world, the organisms within them soon used up almost all the oxygen and today the air trapped above the water in the cave is very low in oxygen but high in methane, carbon dioxide and hydrogen sulphide. No organic nutrients enter the cave from outside. Although volcanic water containing dissolved hydrogen sulphide seeps through the cave on its way to the Black Sea, it comes from an underground reservoir formed thousands of years ago (unlike the ground water in the rest of Romania it contains no traces of radioactivity). Yet the cave contains a flourishing ecosystem. This unique world is supported by bacterial films that coat the cave walls with slime and form scummy mats on the surface of the water. The bacteria digest the limestone walls of the cave to obtain a carbon supply and derive their energy supply by oxidizing hydrogen sulphide. They support a bizarre collection of invertebrates – translucent spiders, millipedes, woodlice, leeches and earthworms. Woodlice and snails eat the microbial mats, and they in turn are preyed upon by spiders and leeches.

The Movile caves can only be reached by diving through flooded passages, but similar, more accessible, sulphur-based ecosystems are found elsewhere. In Southern Mexico lies the Cueva de la Villa Luz, a labyrinth of passages and caves that ramify through limestone rocks. Springs bubble up through the cave floor laden with dissolved hydrogen sulphide and limestone, creating milky pools. Hydrogen sulphide fills the air with the stench of rotten eggs. It condenses on the walls of the cave to form sulphuric acid, which dissolves the rock and burns the skin of the unwary visitor who touches the wall. Despite this seemingly hostile environment, the cave teams with life. Bacterial slime and mucus coat the rocks and drip in gelatinous strings from the ceiling, forming living wobbling stalactites dubbed 'snotties'. Fish swarm in the shallow milky pools, spiders scurry over the rocks, hordes of midges dance in the air. Like the Movile cave, this ecosystem is based on chemosynthetic bacteria that etch away the walls of the cave.

Life without Oxygen

Few multicellular animals can survive without oxygen. But many archaea and bacteria are not only able to live without oxygen, they actually find it so toxic that they cannot tolerate even a brief exposure to the gas and are obliged to live in an oxygen-free atmosphere. Such anaerobic environments are plentiful. They can be found in the ooze that covers the floor of lakes and oceans, in bogs, sewage treatment works, and even in the guts of animals. Some of these organisms use hydrogen as an energy source and carbon dioxide as a carbon source for growth, and produce large amounts of methane in the process. Consequently, they are known as methanogens. Many of them are spherical archaea belonging to the *Methanococcus* family. The ability of cows to eat grass is not innate; it is dependent on symbiotic *Methanococci* in their gut, that break down cellulose. The methane that they produce in the process makes a major contribution to global warming because, like carbon dioxide, it acts as a greenhouse gas.

Although oxygen is abundant in the atmosphere today, this was not always the case. The early atmosphere contained little or no oxygen and consisted principally of carbon dioxide and nitrogen. Oxygen was a waste product produced by photosynthetic single-celled organisms, the cyanobacteria, that evolved around 3 billion years ago, at a time when life was already well established (the first single cells are thought to have arisen around 3.8 billion years ago). These cyanobacteria used the energy of sunlight to convert water and carbon dioxide into carbohydrates. In the process they produced oxygen as a by-product and so created the atmosphere of today's Earth. They also altered the composition of the ocean. The early seas contained large quantities of iron and the oxygen produced by the cyanobacteria was initially consumed by the oxidation of dissolved iron. This precipitated out of solution and formed a band of iron oxide in the ocean floor that began about 2.8 billion years ago and has been used to date the rise of the cyanobacteria. Around 0.5 billion years later, the supply of iron was exhausted and the oxygen level in the atmosphere began to increase, reaching its present level about 0.8 billion years ago. It is salutary to think that a single-celled organism is responsible for what is probably the largest-scale pollution known.

Oxygen was toxic (and still is) to most life-forms, and many died out as the oxygen concentration in the atmosphere gradually rose. Those that survived evolved strategies to protect themselves against the highly reactive oxygen ion. It is a paradox that oxygen, which is essential not only for human survival but for almost all life on our planet, is also a deadly poison. Oxygen is used by intracellular organelles known as mitochondria to make the chemical energy that powers our cells. However, sometimes oxygen grabs an extra electron to become a 'free radical'. These are highly reactive and dash about the cell causing mayhem, because their additional electron needs a partner and will snatch one from any molecule close by. Membranes, proteins, lipids,

DNA – all are fair game to a free radical. The result is a chain reaction, for although the original free radical is stabilized by the stealth of an electron, it creates a new one in the process. Many molecules may be damaged in this way before cellular defence mechanisms are finally able to destroy the errant free radicals; in fact, free radicals are one of the major causes of cell death. Oxidation – the ability of oxygen to rip electrons off other molecules – also makes iron rust, fires burn and fats go rancid.

Oxygen was discovered by Joseph Priestley (1733–1804) while examining the gas given off when mercuric oxide was heated. He observed that 'a candle burned in this air with a remarkably vigorous flame'. He also explored the effects of the gas on mice, by placing them in a small bell-jar filled with a supply of gas; he found that whereas a mouse exposed to ordinary air expired within a quarter of an hour, that exposed to 'pure air', as he called it, was still alive more than half an hour later. Priestley disclosed his findings to the French chemist Antoine Lavoisier (1743–94), who later named the gas oxygen. The name is taken from the Greek and means 'formerly acid' because Lavoisier, mistakenly as it turns out, believed that it was a component of all acids. Unfortunately for science (and for himself), Lavoiser met his death early, at the hands of Madame Guillotine.

Priestley was very perceptive and anticipated the use of oxygen to sustain life. He argued that it might be used to 'agreeably qualify the noxious air of a room in which much company should be confined . . . so that from being offensive and unwholesome, it would almost instantly become sweet and wholesome.' He also conjectured that oxygen might be 'peculiarly salutary to the lungs in certain morbid cases when the common air would not be sufficient'. Early scientists often experimented on themselves and Priestley was no exception. He found that breathing oxygen had no adverse effects and queried whether 'this pure air may become a fashionable article in luxury'. Today, canned oxygen is sold on the streets of Tokyo to give a quick lift to those commuters who are overcome by the city's toxic smog.

Breathing pure oxygen in quantity, however, can be hazardous. In the 1950s, premature infants were given pure oxygen to breathe in the belief that it would help their survival. Unfortunately, the high oxygen concentration in the incubator caused constriction of the fine blood vessels in the eye. As a result, these children developed fibrous tissue behind their eyes, and became blind. There is no danger, however, if the oxygen concentration is kept below 40 per cent. Pure oxygen is still used sometimes by divers and astronauts but special precautions must be taken, as described in Chapters Two and Six.

Cool Characters

Unlike heat, many animals, including humans, can tolerate severe cold. Their adaptations are considered in Chapter Four. Here we consider the extremophiles – those organisms that live in near-freezing conditions and those that are able to tolerate freezing.

Cold *per se* does not damage proteins; it merely slows the rate at which biochemical reactions work. Consequently, most organisms cease to reproduce or even grow (in the strictest sense) a few degrees below zero centigrade. Metabolic activity continues, albeit at a reduced rate, and has been documented for Antarctic lichens at temperatures as low as -27°C. Around -80°C metabolic activity probably stops completely, and the organism then exists in a dormant state. Many cells, including those of humans, can be stored for long periods at the temperature of liquid nitrogen (-196°C). The lowest temperature to which cells can be cooled and survive rewarming is unknown, but is likely to be even lower. Great care is required to cool animals and cells below zero degrees Centigrade, however, because although cold is not itself deleterious, freezing is a very different matter.

Psychrophiles are cold-loving organisms that live in near-freezing water. They are found in the ocean depths, where the temperature is a relatively constant 1–3°C and they live within and below the polar icecaps. They even grow happily in domestic refrigerators. Whole communities of psychrophiles inhabit the Antarctic sea ice, living in thin layers of non-frozen water within the ice. They include a plethora of bacteria, archaea, algae and diatom species, like the snow algae *Chlamydomonas nivalis*, which colours snowfields in shades of soft pink and bright green, and the bacterium *Polaromonas vacuolata* which is distinguished by a preference for a temperature of 4°C, and ceases reproducing if it heats up above 12°C. Nor are such communities devoid of multicellular life. Cruising along the ocean floor in a submersible at a depth of around 550 metres, Charles Fisher spotted a strange multicoloured mushroom-like structure, 2 metres in diameter, sprouting from the bottom. It was crawling with inch-long worms. Investigation revealed that the structure consisted of an ice-like mixture of water and methane (the gas had seeped out of vents on the ocean floor). A flourishing community of bacteria and archaea that feed on methane provided sustenance for the worms.

Deep beneath the Antarctic icecap lie many freshwater lakes, their waters prevented from freezing by geothermal heating. The largest is Lake Vostok, which lies about 4 kilometres below the surface of the ice and is estimated to be 200 kilometres long, 50 kilometres wide and 500 metres deep – roughly the size of Lake Ontario and about twice as deep. Icecaps began to close over Antarctica around 40 million years ago so any life that exists in Lake Vostok has probably been sealed away for several million years, making the lake a time capsule that may contain unique microorganisms that hold information about the planet's history. But scientists' eagerness to explore these icy subterranean lakes is tempered by the difficulty of sampling the water without contaminating it with life from the surface. Such fears stopped an ice-core drilling programme in 1966 just 150 metres before the probe broke through into Lake Vostok. Present-day researchers are still discussing how best to deal with the problem.

Cold is a great preservative because biochemical reactions are dramatically slowed. The cold dry air of Antarctica means that the goods Captain R.F. Scott and his party left in their hut in 1904 are still completely fresh. Mammoths have been found deep frozen in the Arctic, their carcasses so perfectly preserved that their meat is still edible more than 30,000 years after their death. Such frozen tissues provide a valuable historical and biological archive. The reason for their perfect state is that the bacteria that decompose flesh and food simply cannot grow at very cold temperatures, because of the lack of liquid water.

Life in the Freezer

As every gardener knows, freezing is lethal for many plants. Late spring frosts nip developing blossom in the bud and the first hard frost of winter can turn a lush border of summer flowers into a shrivelled brown mush. Most animals also cannot tolerate freezing.

The investigation into the effect of freezing on life has a long history. Around 1663, Henry Power observed that when he placed a jar of vinegar containing 'minute eels' in a mixture of ice and salt, the liquid froze and the eels were 'incrystalled'. But when the frozen vinegar was allowed to thaw, the eels once again 'danced and frisked about as lively as ever'. Robert Boyle was also fascinated by the effects of freezing and attempted to freeze frogs and fish – with limited success. The first experiments on insects were carried out by Réaumur, a French scientist who built one of the first thermometers and was therefore able to quantify his observations. He observed that a common species of caterpillar survived freezing to -20 °C, while another unnamed species was only able to tolerate -11 °C. He also discovered that their blood froze at different temperatures, likening them to brandies of different strengths because strong spirits froze less readily than weaker ones. This was the first suggestion that freeze tolerance might depend upon specific physico-chemical properties of insect blood, and anticipated recent studies that have identified the natural antifreezes involved.

The advent of the golden age of mountain and polar exploration brought with it many more fabulous tales of freezing and resurrection. One of the most bizzare was reported by Turner in 1886, who described how dogs fed with Alaskan blackfish that had been chopped out of blocks of ice vomited up living fish shortly afterwards. The warmth of the stomach had unfrozen the fish and brought them back to life. While it may be hard to credit this story, no one would question the reliability of the British explorer John Franklin. During a voyage to the northern polar seas, he recorded that a carp that had been frozen for thirty-six hours leapt about wildly when it was thawed out. Despite these travellers' tales, however, freezing is lethal for most cells.

Frost damage occurs because ice crystals form within and between cells. Razor-sharp ice needles puncture the delicate membrane that surrounds each cell, which allows the cell contents to leak out. Intracellular membranes that divide the cell into discrete compartments are also ripped apart, so that the contents of the organelles mix and biochemical reactions are disrupted. Ice is a crystal of pure water but biological solutions contain many salts. When ice forms in the extracellular solution, therefore, the salt concentration in the solution that remains unfrozen increases. This creates an osmotic force that drags water out of cells, causing them to shrink and increasing their internal salt concentration. Ice formation inside the cell increases the salt concentration of the intracellular solution directly. The resulting dehydration damages the cell membrane and cellular proteins. Freezing may also disrupt the connections between cells and damage the capillaries that supply them, leading to oxygen and nutrient starvation. As described in Chapter Four, frostbite can cause severe damage in humans. Nevertheless, some plants and animals are not harmed by freezing temperatures.

Freeze-tolerant organisms use two strategies to combat cold. Some lower the temperature at which ice crystal formation takes place by synthesizing natural antifreezes; remarkably, others simply freeze solid.

The blood of many insects and fish contains natural antifreeze substances which prevent freezing of the body fluids at temperatures below zero (a phenomenon known as supercooling). The winter flounder *Pseudopleuronectes americanus*, for example, synthesizes at least seven different antifreeze proteins when the temperature falls to around 4 °C. The common yellow mealworm *Tenebrio mollitor*, used as bait by fishermen, contains an even more potent antifreeze. Antifreeze proteins lower the freezing point of water by binding to the surface of developing ice crystals and inhibiting their growth. They have no effect on the melting point of ice that has already formed. Some insects that supercool to even lower temperatures use low molecular weight alcohols, such as glycerol, as antifreezes. These work on the same principle as the ethylene glycol that is added to car radiators in winter in Northern Europe to prevent the cooling water from freezing. As much as 20 per cent of the body fluid of the gall moth (*Epiblema scudderiana*) can be glycerol, which enables the insect to supercool to -38 °C without freezing.

Supercooling can be risky, however. If the temperature dips below the supercooling point, the tissues freeze instantly. This can be lethal. Flash freezing can be precipitated by ice crystals propagating across the skin or by contact with ice-nucleating agents, which provide a base around which ice crystals form (as may occur if the skin is damaged). Some moths and butterflies surround themselves with protective silk cocoons to prevent their skin coming into direct contact with ice.

Other animals adopt an alternative strategy and freeze solid over winter. The woolly bear caterpillar (*Gynaephora groenlandica*), which lives in the high Arctic, spends the greater part of the year – often as much as ten months – frozen solid at temperatures of -50 °C or lower. The Siberian salamander, *Salamadrella keyserlingii*, is equally remarkable. It lives high in the Arctic Circle where all but the upper few metres of the soil are permanently frozen and the surface layers also freeze in winter. During the brief Arctic summer the adult salamanders run around actively, and lay their eggs in the shallow ponds and puddles that lie scattered over the tundra. In winter, they hibernate in moss cushions close to the ponds, where the temperature can drop to -35°C. They have been found frozen solid in ice as deep as 14 metres below the surface of the tundra. Yet when spring comes and the tundra thaws, the salamanders simply thaw out, get up and run off. Newly hatched painted turtles, garter snakes and many species of frog also freeze solid in winter. Zoologists trying to understand how they do so must keep their chosen charges in the freezer.

To endure freezing, ice crystals must be kept small to ensure that they do not puncture cell membranes. This is achieved by specialized proteins that act as ice-nucleating agents, which are synthesized as the temperature drops in autumn. These proteins seed ice crystal formation, creating thousands of tiny ice crystals in the extracellular fluids. Small crystals have a tendency to coalesce into larger ones over time, as can be observed in ice cream that is frozen for long periods. To prevent this recrystallization, animals use additional antifreeze proteins that stabilize the small and harmless ice crystals and prevent their enlargement. Freezing is thus a slow, controlled process that enables the cell to adjust gradually to change.

Another serious problem for creatures that freeze solid is that cell water is lost and the cell shrinks when the extracellular fluid freezes. This may denature the cell membrane and damage cellular proteins, and freezing of more than 65 per cent of the body water is usually lethal. Freeze-tolerant animals prevent these changes in cell volume by increasing the concentration of sugars or amino acids within their cells. These cryoprotective substances reduce ice formation, decrease cellular water loss, and stabilize the cell membrane so that it is able to withstand greater shrinkage without damage. They include glycerol, and sugars such as trehalose (in insects) and glucose (in frogs).

Like freezing, thawing is a controlled process. When frozen frogs thaw out, for example, the heart melts first, enabling the recovery of vital functions to begin at once and speeding the thawing process by facilitating the circulation of warm blood.

Suspended Animation

Recent advances in technology have enabled mammalian cells and tissues to be cooled to very low temperatures with comparative ease.

Toads are good diggers and pass the winter beneath the frost line, safe in their burrows. Frogs are unable to dig and so they hibernate in the leaf litter of the forest floor where the temperature falls as low as -8°C. The wood frog shown above freezes solid – 65 per cent of its body water is ice. Specialized proteins ensure that the ice crystals remain too small to cause any damage. The frog's vital organs are prevented from freezing because large quantities of glucose are produced by the liver, creating a highly concentrated sugar solution that permeates the tissues and acts as an antifreeze.

Cryopreservation takes its name from the Greek *kruos,* meaning icy cold. The low temperature slows cellular metabolism so much that it produces a state of suspended animation and enables cells to be preserved for far longer than their natural life-span. The colder the temperature, the slower the metabolic rate, and thus the longer the cells can be maintained. To limit damage from dehydration or ice crystal formation, the rate of freezing and thawing must be carefully controlled, and cryoprotectants must be added to the solution in which the cells are suspended. Glycerol is commonly used as a cryoprotectant because it can prevent water being converted to ice even at liquid nitrogen temperatures.

Sperm is routinely frozen in liquid nitrogen, at a temperature of less than -196 °C, for artificial insemination. Originally developed for cattle,

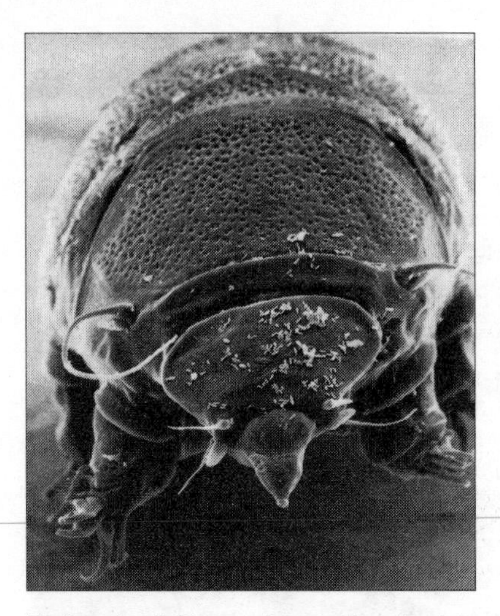

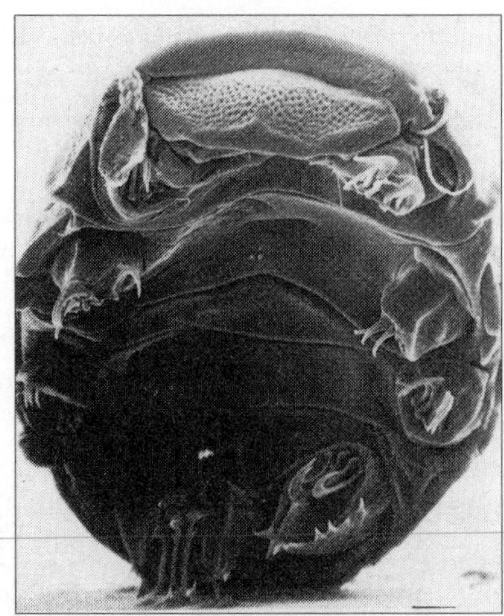

Tardigrades (left) are microscopic animals, around a millimetre in length, that live in damp sand, the ooze at the bottom of lakes and the depths of the sea, and the thin film of water that coats the leaves of moss in the Arctic tundra. More prosaically, they can be found in the mossy pincushions that sometimes clog the rainwater gutters of your house. Nicknamed waterbears because of their stumpy clawed feet and lumbering gait, their most remarkable feature is their ability to survive extreme conditions in a state of suspended animation. When times get tough, the tardigrade curls up, draws in its legs, and turns into a resting stage known as a tun (right), in which metabolism almost stops. Water loss is dramatically reduced, and the accumulation of trehalose and ice-nucleating proteins helps protect the tun from the effects of desiccation and extreme cold. The tun is the ultimate survivor. It can withstand temperatures as cold as -272°C (only one degree above absolute zero) or as hot as 151°C, extreme drought, immersion in liquid alcohol (which pickles most creatures), and pressures of around 6000 atmospheres (most organisms, including bacteria, are killed by pressures of more than 3000 atmospheres). But just add water, and the tun rehydrates to form a tardigrade that seems quite unconcerned by its experience. They have even been revived from samples of a dried-out moss that had been stored for 120 years in an Italian museum.

this technique was first successfully applied to human sperm in 1953. Frozen semen samples can retain their potency for decades and human sperm stored for more than fifteen years has resulted in conception. Many men elect to freeze their sperm prior to having a vasectomy, or undergoing chemotherapy or radiation treatment for cancer, which are known to damage sperm. Others provide semen samples for infertile couples. Thousands of babies are born each year from cryopreserved semen samples, and sperm that are frozen and then thawed before insemination are thought to be no more likely to result in birth defects than fresh sperm.

Embryos can also be preserved by freezing them at low temperatures. First developed in domestic animals, this technique is now routinely used as part of the normal process of *in vitro* fertilization (IVF) in humans. Usually, multiple eggs are retrieved from a woman in a single operation, fertilized *in vitro* and then two or three of the embryos that develop are transferred back into her womb. Any embryos that are not used are frozen, just in case the first embryos that were implanted do not result in pregnancy. This avoids women having to undergo the stress of repeated egg retrieval and, because this is the most expensive part of the operation, reduces the cost of subsequent IVF trials. Spare embryos can also be stored for several years, in case the couple decide to have more children at a later date, or where the woman has to undergo medical treatment that may affect her fertility, or even for use by women who are unable to produce their own eggs. The first child to be born from a frozen embryo was Zoe Leyland, as recently as 28 March 1984, in Melbourne, Australia. Successful pregnancies have now been achieved with embryos that have been frozen for as long as five years.

Other types of human cells can also be frozen. Undoubtedly, the most famous are HeLa cells. These were isolated from a tumour in a patient called Henrietta Lacks (hence HeLa) and frozen immediately in liquid nitrogen. Many years after her death, offspring of the original tumour cells can be found in research laboratories throughout the world and provide a valuable resource for medical research.

Although mammalian cells can be frozen and thawed with relative impunity, this is not the case for the whole animal. Nevertheless, in the United States today, there are several cryopreservation companies that freeze the bodies (or heads) of the recently dead in the hope that they can be reawakened by future generations, their ills cured, their worn-out parts replaced and a further period of useful life assured. Most of these companies are located in California, where the legal system is most amenable to cryogenic preservation. Sadly, the dreams of their clients face many difficulties, for after death the body's tissues are quickly damaged by the lack of blood flow.

There is one sense, however, in which freezing can preserve an individual. Their individual genetic constitution – their genes – will survive. But for this, all that is needed is a few cells, which a simple blood sample can provide (although human red blood cells have no nucleus, and thus no genetic information (DNA), there are enough white blood cells to provide all the DNA required). One day it may be possible to produce a human from a single white blood cell, using the same type of technology as that which generated the famous sheep clone, Dolly. Whether we will want to do so is another matter. It is worth remembering that even if it were possible to clone one of your own cells, the resulting person would be no more similar to you than your identical twin. You are far more than just your genes.

Making Millions from Microbes

Extremophiles are becoming big business. A growing industry is based upon enzymes isolated from organisms that can tolerate extremes of heat, cold, salt, acid, pressure and heavy metals, to name but a few. Small biotechnology research companies send their employees to the ends of the Earth to find new extremophiles that may possess previously unknown genes, and then race to patent any that are discovered. The competition is intense, for the potential rewards are high.

Thousands of molecular biologists exploit the power of hyperthermophiles in their research every day. Heat-tolerant enzymes are used to make multiple copies of a selected piece of DNA in a process known as the polymerase chain reaction (PCR). As its name suggests, this process involves successive reaction cycles. First, DNA must be heated to break it apart into its two constituent strands. It is then cooled, and each strand is replicated with the help of an enzyme. These two steps are repeated many times, which results in an exponential increase in the number of DNA molecules. The PCR technique relies on a DNA-replicating enzyme that does not fall apart at the high temperatures (95 °C) needed to separate the two strands of DNA. Fortunately, hyperthermophilic enzymes, such as *Taq* polymerase, have evolved to do just that. The PCR method is not confined to research laboratories. It is widely used in medicine: for example, to identify bacterial strains or for genetic testing of individuals. And it has revolutionized forensic science because the technique is so sensitive that it is possible to make billions of copies of DNA from a few molecules, and thus to identify a criminal from one stray cell inadvertently left behind at the scene of the crime.

Taq polymerase was isolated from the archaea *Thermophilus aquaticus* discovered in the superheated bubbling pools of Yellowstone National Park by Thomas Brock. It languished in the laboratory for over twenty years until Kary Mullis realized that it could be used to make multiple copies of DNA. A brilliant but colourful character, Mullis offended many members of the scientific establishment with his flamboyant statements and unconventional lectures punctuated with slides of surfing or of his girlfriends in compromising positions. Nevertheless, he received the Nobel prize – and deservedly so, for his work has transformed the life sciences and the PCR technique that he invented has become the workhorse of modern molecular biology. *Taq* polymerase was the first enzyme extracted from an extremophile to be commercially exploited and its sales generate an income of over 80 million dollars a year. Companies are still fighting over the patent.

Enzymes extracted from alkaliphiles are much sought after by manufacturers of washing powders. Enzymes are added to biological washing powders to help break down the protein, sugars and fats that cling to dirty garments. But detergents are highly alkaline, and most enzymes cannot operate effectively under these conditions. Those of alkaliphiles, however, function best at high pH. In 1997, the United States company Genecor introduced a detergent containing an enzyme extracted from an alkaliphile discovered in a caustic lake, that is claimed to enable clothes to be washed hundreds of times and still look like new. It works by breaking down the fine surface layer of fluff that dirt collects on, without affecting the underlying fabric. This was the first large-scale industrial application of a product from an extremophile.

There are a host of other potential applications for extremophiles. Acidophiles can facilitate the recovery of valuable metals from low-grade ores, a process known as microbial leaching which is becoming increasingly popular for extraction of gold, copper and uranium. Enzymes extracted from psychrophiles could be used for cold-water soaps and detergents, and for catalysing reactions that must be carried out in the cold. Bacteria and archaea are increasingly being exploited in bioremediation projects, and to degrade toxic compounds, such as pesticides, petroleum and solvents. The commercial exploitation of extremophiles may still be in its infancy, but the potential applications are enormous.

Life Beyond Earth?

In August 1996, a small, rather ordinary-looking lump of rock called ALH84001 made headline news.[5] While most scientific articles are lucky to be read by more than a handful of aficionados, the findings of this paper were discussed in detail in newspapers and on radio and TV news bulletins throughout the world, even before the article was published. The excitement the paper generated is understandable. NASA scientists claimed to have found evidence for life on Mars.

Sixteen billion years ago, a meteorite impact on Mars chipped many small fragments of rock off the surface of the planet, blasting them into space. Around 11,000 years ago, one of these rock fragments was captured by the Earth's gravitational field and plummeted to rest in the Allen-Hills ice field of Antarctica. ALH84001 is thus a visitor from Mars, one of a small number of such meteorites that have been discovered. Their Martian origin is proven both by the mineral content of the rock and by the composition of gas bubbles sealed within it. These match those of the surface rocks of Mars and of the Martian atmosphere, which were measured by the Viking probe that reached the red planet in 1976.

Deep inside ALH84001, the scientists discovered irregularly shaped structures that resembled those of terrestrial microfossils, formed nearly 4 billion years ago. Additional data led them to suggest that while each piece of evidence might have a different explanation when considered in isolation, when taken together they provided 'evidence for primitive life on early Mars'. Alas, their conclusions were probably premature. Excited by the possibility of extraterrestrial life, several international teams of scientists turned their attention to ALH84001, analysing and reanalysing the same lump of rock. After a year's work, the general consensus was that the structures observed were simply mineral deposits, rather than the fossil remains of alien life-forms.

But the possibility of life elsewhere in the solar system cannot be abandoned easily. The extreme conditions that many archaea can tolerate resemble those found on other planets or their moons. The cold dry rock valleys of Antarctica are so close to the conditions encountered on Mars that they are used to test instruments designed for missions to the red planet. Yet inside the rock, about a millimetre below the surface, lives a thin layer of photosynthetic microorganisms.

Unbelievably, bacteria can even survive the vacuum of space. The Surveyor 3 probe landed on the moon in April 1967. Two and a half years later, it was visited by astronauts on the Apollo 12 mission whose aim was to investigate how the probe had withstood the severe conditions – intense solar radiation, a near vacuum and extreme temperature shifts. They removed a TV camera and took it back to Earth in a sealed container, where it was opened under sterile conditions in the lunar receiving laboratory. Microbiologists cultured samples taken from inside the camera and found, to their amazement, that microorganisms grew. But these were not novel moon bugs – they were familiar Earth varieties. It seems that during the manufacture of the TV camera, a technician had sneezed and some bacteria had landed deep inside the instrument and been sealed there until the camera was reopened in the lunar laboratory. Of course, the sceptic might argue that the bugs never made it to the moon at all, but contaminated the camera after its return to Earth. However, the stringent sterile conditions employed during recovery and analysis of the camera make this unlikely. It appears that the bacteria were indeed able to survive two and a half years on the surface of the moon.

There is a difference, however, between survival and growth. Without liquid water, life (at least as we understand it) cannot exist except in a

High-resolution image of Europa, obtained by the Galileo spacecraft. Europa, one of Jupiter's sixteen moons, was discovered by Galileo in 1610. It is unique in the solar system in that it has a smooth surface with relatively few craters or mountains. Its crust is criss-crossed by an intricate web of dark lines that are believed to be fractures in the outer layer of ice that cloaks the satellite.

state of suspended animation. Growth and reproduction are simply not possible. So the search for life elsewhere in the solar system is really the search for water. And there are places where liquid water may exist. In 1979, the Voyager probe reached Jupiter and discovered that its satellite Europa was covered with a layer of ice. More recent data obtained by the Galileo spacecraft suggests that an ocean of liquid water may lie trapped many miles beneath the moon's frozen crust, like the large lakes that exist beneath the surface of Antarctica. Scientists are currently planning to send a further spacecraft to explore this possibility and to investigate if life exists on Europa. It will be exciting to find out.

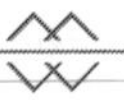

NOTES

1. LIFE AT THE TOP

1 W.J. Turner (1889–1946) 'Romance'.

2 Mallory's famous reply to the question, 'Why do you want to climb Mount Everest?'

3 They were warned of this problem by Paul Bert, but his letter arrived too late, for the date of the flight was already fixed. They decided to go ahead anyway.

4 The precise CO_2 concentration in the atmosphere has always been contentious. In the early part of the twentieth century, it was variously placed at 0.04 per cent and 0.033 per cent and J.S. Haldane experimented on the roof of the Physiological Laboratory in Oxford, trying to obtain a precise value. Today, the issue is whether the CO_2 level is rising as a result of the use of fossil fuels. Interestingly, the CO_2 level in the atmosphere may vary throughout the globe. When the temperature falls below -70 °C, as can happen in Antarctica, CO_2 freezes and its concentration plunges to zero. This phenomenon is even more extreme on Mars, where the atmosphere is almost entirely CO_2 and thins during winter when CO_2 freezes, only to be reconstituted in the spring when the temperature rises and the solid gas vaporizes.

5 The technical term is ventilation, which is defined as the volume of air breathed in (or out) each minute. The average man inspires around half a litre of air in each breath and takes around twelve breaths each minute, so his ventilation rate is six litres/min. The maximum possible rate is about 150 litres/min (but this can only be achieved by top athletes).

6 Strictly speaking, the carotid bodies measure the partial pressure of oxygen in the blood. In physiology, the terms used to describe the oxygen concentration in the blood have very precise meanings (for very good reasons). The partial pressure of oxygen in the blood refers to the partial pressure of the dissolved gas. Oxygen content – the total amount of oxygen in the blood – is approximately equal to the oxygen bound to haemo-globin (as so little is actually dissolved in solution). It therefore depends on the number of red blood cells and increases when the haematocrit rises. Oxygen saturation describes the percentage of haemoglobin that has bound oxygen.
7 Acclimatization can be speeded by the drug acetazolamide, which works by stimulating the kidneys to excrete bicarbonate ions and return the blood acidity to normal. It also helps keep carbon dioxide levels high around the central chemoreceptors. Acetazolamide not only speeds acclimatization, it also ameliorates acute mountain sickness.
8 Recent studies by Luke Howard and Peter Robbins offer new insight into the processes underlying the initial changes in breathing in response to altitude. In their Oxford laboratory, they showed that ventilation increases at low atmospheric oxygen levels even when the blood acidity is kept constant by carefully adjusting the carbon dioxide level in the inspired air. This suggests that low oxygen *per se* may have a more important effect on breathing than previously envisaged. The mechanism that underlies this phenomenon is unknown, but it has been suggested that it involves an increase in the sensitivity of the carotid bodies.
9 The barometric pressure is higher near the equator because of a large mass of cold air in the atmosphere above the equator, which in effect presses down on the air below.
10 It is perhaps worth noting that Mabel herself did not do this.

2. LIFE UNDER PRESSURE

1 The Kraken is a mythical sea monster of enormous size said to live off the coast of Norway. It is immortalized in a poem of that name by Alfred Lord Tennyson.
2 You only pay on leaving.
3 One of the first to describe the phenomenon was Robert Boyle, who

in 1670 observed how a bubble formed in the eye of a viper on decompression.

4 J.B.S. Haldane later related that this was a far from comfortable experience. His diving dress ended at the wrists in rubber cuffs to keep out the water. Because of his small size, the cuffs leaked and water flooded the diving dress right up to his neck. Fortunately, the air that was pumped down to him prevented the water from rising any higher, but he got very cold indeed.

5 My father once repeated to me a rhyme he was told by the famous British diver Buster Crabbe: 'Down at a depth of 30 feet, lies a devil by the name of Oxygen Pete.' The name Oxygen Pete was coined when a naval rating regained consciousness after a fit induced by oxygen. When he inquired who had knocked him out, he was told it was Oxygen Pete. The name caught on.

6 Cave divers sometimes use pure oxygen, as the smaller cylinder is an advantage when trying to squeeze through narrow holes.

7 These men were all members of the International Brigade, Communists who fought against Franco in the Spanish Civil War, a fact which gave Haldane great satisfaction (he was a strong supporter of Communism at the time). He wrote that he 'chose these men as colleagues because I had no doubt of their courage and devotion', rationalizing that men with battle experience were likely to remain cool under pressure.

3. LIFE IN THE HOT ZONE

1 One calorie is the amount of energy required to raise the temperature of 1 gram of water by 1 °C. Because this quantity varies slightly with temperature and pressure, it is more precisely defined as the energy required to raise the temperature of one gram of water from 15 °C to 16 °C. It is exactly one thousandth of the Calorie (note the capital C) used to calculate the nutritional content of food, which is more correctly referred to as a kilocalorie. The energy used to evaporate water is liberated as heat when steam condenses, which is why steam causes a much more severe burn than water at the same temperature.

2 Les A. Murray 'A Retrospect of Humidity.'

3 The scientific name of Ecstasy is 3,4-methylene

dideoxymethamphetamine. It is an amphetamine derivative.

4 This argument applies only to adults. Young children are susceptible to febrile convulsions and cooling is advisable.

5 When working in a hot climate, as much as 18 litres of water a day may be needed to prevent dehydration: an amount equivalent to 36 cans of Coke.

4. LIFE IN THE COLD

1 Weight for weight, fat contains more calories than protein or carbohydrate, so their diet was predominantly based on fat: 57 per cent fat, 35 per cent carbohydrate and 8 per cent protein. They even drank their hot chocolate with butter. This may also be one reason that Tibetan monks in Himalayan monasteries drink their tea with yak butter (a singularly unappetizing taste to the Western palate).

2 Hypothermia can cause bizarre behaviour. One Channel swimmer who asked for a cloth to wipe his eyes, ate it instead. Another was convinced she was being chased by furry animals.

3 His story, and that of the tragedy on Everest that May is told in *Into Thin Air,* a gripping account by Jon Krakaeur. The quotations come from this book.

5. LIFE IN THE FAST LANE

1 These values are for male athletes; females use less. The Calorie (note the capital C) used to calculate the nutritional content of food is more correctly referred to as a kilocalorie because it is exactly one thousand times larger than the calorie, the unit used to measure heat. One calorie is the amount of energy required to raise the temperature of one gram of water by one degree Centigrade, so that a daily intake of 2000 kcal is roughly equivalent to the amount of heat needed to heat 20 litres of water from 0 °C to boiling point. It is no wonder you get hot when you run.

2 At the 1908 Olympic Games in London, the marathon was run from Windsor Castle to White City, a distance of 26 miles. This is the origin of the length of the modern marathon.

3 The heart rate decreases with age. You can calculate your own

maximum heart rate quite simply, by subtracting your age from 220.

4 In the ancient Olympics, married women were not only not permitted to compete, they were also banned from attending on pain of death. Callipateira, so Pausanias tells us, disguised herself as a gymnastic trainer, so great was her desire to see her son compete. She was discovered, but out of respect for her father, her brothers and her son, all of whom had been victorious, she was not punished. But the Greeks passed a law stipulating that all trainers must strip before entering the arena, so that her crime could not be copied.

5 The admission by former East German athletes and trainers that they used performance-enhancing drugs has had some far-reaching repercussions. In 1998, four US women swimmers, who were beaten into second place in the relay race at the 1976 Montreal Olympics by the East Germans, requested that their silver medals be upgraded to gold. Britain's Sharon Davies, who was narrowly beaten by Petra Schneider in the 1980 Olympics, also asked for the records to be reviewed.

6. THE FINAL FRONTIER

1 A speed of around 40,000 kilometres per hour is required to achieve orbit, the precise figure depending on whether it is to be a low or high orbit.

2 Your speed is 1670 km/hour at the equator. In the UK, it is only 1075 km/hour because the Earth's circumference is smaller, and at the North Pole it would be almost nothing.

3 In the past, radiation from natural sources was excluded from legislation (it was, after all, natural) and levels were only advisory. But new legislation to be implemented in the European Union will include cosmic radiation received by flight crew as an occupational exposure and will require various protection measures (all of which are already fulfilled by UK airlines). The risk of a fatal cancer with a radiation dose of 1 msv (the current annual recommended limit for the general public) is 1 in 20,000. For people who choose to work in occupations associated with radiation risk, the recommended limit is 20 msv, equivalent to an annual risk of death due to irradiation of 1 in 1000.

4 'High Flight' by John Gillespie Magee. Magee was a pilot with the Royal Canadian Air Force during the Second World War. He began

this sonnet while flying at 9000 metres and finished it soon after he landed. He died shortly afterwards, aged just nineteen.

5 Car tyres work in a similar way. They are strengthened with wires to prevent them exploding, because the pressure inside the tyre can be six times that of the outside air.

6 The body fluids decrease by roughly 0.8 litres in space, so this amount should restore the volume back to that on Earth. The saline solution astronauts drink is similar to the oral rehydration solutions given to Earth-bound individuals who have lost fluid as a result of vomiting or diarrhoea.

7 They left the memorial surreptitiously and did not discuss it until their return to Earth.

7. THE OUTER LIMITS

1 *Deinococcus radiodurans* seems to have developed resistance to radioactivity as a fortuitous side-effect of resistance to drought. It is able to reconstruct its chromosomes even when radiation has split them apart into a myriad pieces. How it achieves this feat remains something of a mystery.

2 At a depth of 3000 metres the pressure is so great that water boils at 400 °C.

3 It moves from a highly ordered state to one of complete disorder. This principle is enshrined in the second law of thermodynamics – the law that, in Tom Kirkwood's wonderful phrase, describes 'the tendency for dirty dishes to collect in the sink.'

4 Of course, it may be that the laboratory conditions are not optimal, which would lead to an underestimate of bacterial numbers.

5 It is named after the location and year of its discovery: Allen Hills (19)84.

A NOTE ON THE UNITS

All scientists use a common measurement system. That at least is the goal and, in general, it is true. But it has not always been the case; a colourful variety of units and measurement devices are scattered throughout the texts of earlier scientists. Even in my own lifetime, what appeared to be standard units have metamorphosed into new ones, or changed their names as the committees who set the units favoured one scientist over another. In this book, I have endeavoured to use standard scientific units. These will hold no fears for most Europeans who are accustomed to the metric system. The British, as ever, are an exception (despite the gradual introduction of metric units, my local market still sells fruit in pounds rather than kilograms). Those of you who live in the United States and Canada may also be less familiar with the units in this book. For you, and for reluctant Britons, the next two pages define the scientific units I have used and explain how they may be converted into more familiar ones.

Studies of altitude, and of respiratory physiology, have always been complicated by the variety of units used to describe pressure and height. For simplicity, I have given all altitudes in metres: it is possible to convert these directly to feet by multiplying by 3.28. Pressure has variously been expressed as pounds per square inch, millimetres of mercury (also known as Torr), or, more recently, kiloPascal (kPa). Since much of the early literature and most physiological textbooks use Torr (millimetres of mercury), I have also done so. Younger people, who may be more familiar with kPa, will need to multiply throughout by 0.133.

A bewildering multiplicity of units has also been used to describe depth and pressure underwater. I have given all depths in metres: it is possible to convert these to feet by multiplying by 3.28. Traditionally, depths were measured in fathoms, a unit derived from the span of a man's outstretched arms. A fathom measures 6 feet, or 1.83 metres. Underwater, pressure is normally referred to in units of atmospheric pressure (or bars) and I have also done so here. One bar (one atmosphere) equals 760 Torr or 15 pounds per square inch. In the diving industry, pressure is also described in terms of depth, that is, as metres of sea water (msw): 10 msw is equivalent to 1 bar. At a depth of 30 msw the pressure is 4 bar, being the sum of the pressure at the surface (1 bar) and that underwater (3 bar).

Three different temperature scales are in general use. Two of these are widely known: the Centigrade (or Celsius) scale and the Fahrenheit scale. I have used the Centigrade scale because it is not only the one used by life scientists but also that commonly used throughout Europe. Zero degrees Centigrade is the temperature at which water freezes and 100 °C the temperature at which water boils at sea level. The equivalent temperatures in Fahrenheit are 32 °F and 212 °F. To convert Centigrade into Fahrenheit it is therefore necessary to multiply by 1.8 and then add 32. To convert Fahrenheit into Centigrade, subtract 32 and then divide by 1.8. Physicists use the Kelvin scale in which temperature is expressed relative to absolute zero (-273 °C), the lowest possible temperature. Thus, the temperature in degrees absolute is simply the Celsius temperature plus 273.

The amount of energy deposited in a persons tissues by absorbed radiation (the *absorbed dose*) is measured in Gray units (1 Gy = 1 Joule per kilogram of absorbed ionization). The dose is more usually expressed as the *effective dose*, which is a measure of the exposure of the whole body and is obtained by multiplying the absorbed dose by factors that take account of the different effectiveness of the various types of radiation in damaging individual tissues. This is expressed in Sieverts (Sv). The maximum lifetime limit for NASA space crew is 4 Sv for men and 3 Sv for women. Because the Sievert is a large unit, radiation doses are usually given in milliSieverts (mSv) or microSieverts (μSv). 1 Sv equals 1000 mSv or 1,000,000 μSv. When radiation doses are larger (as for astronauts) the dose is often expressed in Gray rather than Sieverts.

Finally, following the US (rather than the UK) convention, I use billion to refer to 1000 million (10^9).

FURTHER READING

The books and articles listed here fall into two categories. The first are accounts of life at the extremes by those who have been there and experienced it for themselves. These are adventure stories that provide a gripping read. The second kind of book provides additional information about the physiology of humans and other animals, and their ability to cope with extreme conditions, usually in a form that is accessible for the general reader. For those who wish to know still more, details of the many papers and articles on which this book is based can be found at the following website: www.**fire**and**water**.co.uk

GENERAL

Attenborough, D. (1998) *The Life of Birds* BBC Books

Boorstin, D. (1983) *The Discoverers* Penguin Books

Case R.M. and J.M.Waterhouse (1994) *Human Physiology: Age, Stress and the Environment* Oxford University Press

Haldane, J.S. (1922) *Respiration* Yale University Press

Haldane, J.B.S. (1940) *Keeping Cool and Other Essays* Chatto & Windus

Schmidt-Nielsen, K. (1997) *Animal Physiology: Adaptation and Environment* (5th edn) Cambridge University Press

Stroud, M. (1998) *The Survival of the Fittest* Jonathan Cape

1. LIFE AT THE TOP

Hunt, J. (1954) *The Ascent of Everest* Hodder & Stoughton

Messner, R. (1979) *Everest: Expedition to the ultimate* Kaye & Ward

Venables, S. (1989) *Everest. Alone at the summit* Odyssey Books

Venables, S. (1989) *Everest Kangshung Face* Hodder & Stoughton

Ward, M., J.S. Milledge and J.B.West (1995) *High Altitude Medicine*

and Physiology (2nd edn) Chapman & Hall.
West, J.B. (1998) *High Life. A History of High Altitude Physiology and Medicine* Oxford University Press.
Whymper, E. (1891) *Travels amongst the Great Andes of the Equator* John Murray.

2. LIFE UNDER PRESSURE

Beebe, W. (1934) *Half Mile Down* Harcourt Brace and Co.
Case, E.M. and Haldane, J.B.S. (1941) 'Human physiology at high pressure I: Effects of nitrogen, carbon dioxide and cold' *Journal of Hygiene* 41, pp. 225–249
Clark, R. (1968) *J.B.S.: The Life and Work of J.B.S. Haldane* Oxford University Press
Hong, S.K. and H. Rahn (1967) 'The Diving Women of Korea and Japan' *Scientific American,* 216, pp. 34–43
Wells, M. (1998) *Civilization and the Limpet* Perseus Books

3. LIFE IN THE HOT ZONE

Cabanac, M. (1986) 'Keeping a cool head' *NIPS* 1, pp. 41–4
Gasmow, R.I. and J.F. Harris (1973) 'The infra-red receptors of snakes' *Scientific American,* 228, pp. 94–100
Taylor, C.R. (1969) 'The eland and the oryx', *Scientific American,* 220, pp. 88–95
Walker, A. and P. Shipman (1996) *The Wisdom of Bones*. Weidenfeld and Nicolson
Wolf, A. (1958) *Thirst.* Charles C. Thomas

4. LIFE IN THE COLD

Brett-James, A. (1966) *1812*. Macmillan
Bullimore, T. (1997) *Saved* Little, Brown & Company
Cherry-Garrard, A. (1994) *The Worst Journey in the World* Picador
Heinrich, B. and H. Esch (1994) 'Thermoregulation in Bees', *American Scientist* 82, pp. 164–70
Krackauer, J. (1997) *Into Thin Air* Macmillan
Scott, R.F. (1913) *Scott's Last Expedition* Smith Elder
Spindler, K. (1994) *The Man in the Ice* Random House

5. LIFE IN THE FAST LANE

Bannister R. (1954) *First Four Minutes* Putnam

McArdle, W.D., F.I. Katch and V.L. Katch (1994) *Essentials of Exercise Physiology* Lee and Febiger
McGowan, C. (1999) *Diatoms to Dinosaurs The Size and Scale of Living Things* Penguin Books

6. THE FINAL FRONTIER

Chaikin, A. (1998) *A Man on the Moon.* Penguin Books
Beatty, J.K. and A. Chaikin (1981) (eds) *The New Solar System.* Cambridge University Press
Nicogossian, A.E., C.L. Huntoon and S.L. Pool (1994) *Space Physiology and Medicine* Lee & Febiger
Sharman, H. and C. Priest (1993) *Seize the Moment: the Autobiography of Helen Sharman* Victor Gollancz

7. THE OUTER LIMITS

Copley, J. (1999) 'Indestructible' *New Scientist,* 23 October 1999, pp. 45–8
Gross, M. (1998) *Life on the Edge* Plenum Press
Madigan, M.T., J.M. Martinko and J. Parker (2000) *Biology of Microorganisms* (9th edn) Prentice Hall
Pain, S. (1998) 'The Intraterrestrials' *New Scientist,* 7 March 1998, pp. 28–32
Pain, S. (1998) 'Acid House' *New Scientist,* 6 June 1998, pp. 43–6
Pappalardo, R.T., J.W. Head and R. Greeley (1999) 'The Hidden Ocean of Europa' *Scientific American,* 281, pp. 34-43
Storey, K.B. and J.M. Storey (1990) 'Frozen and Alive' *Scientific American,* 263, pp. 62–67

INDEX

Page numbers in **bold** refer to boxes
Page numbers in *italic* refer to illustrations